The Geopolitics of Red Oil

Energy security has emerged as one of the most important contemporary geo-political issues. Access to reliable, cheap energy has become essential to the functioning of modern economies but the uneven distribution of energy supplies has led to perceptions of significant Western vulnerability. At the same time, many in the West have become wary of China's re-emergence as a major power in global politics, with its impact on Western foreign policies and potential threat to Western energy security.

This book offers fresh insights into the rise of China as a global superpower and the ways in which its rise is perceived to threaten Western energy security, engaging specifically with how the idea of the China threat has emerged in popular discourse. The author questions how recent US foreign policy has sought to position China as an antagonist to Western energy interests and explores how this image has become the dominant understanding of China by the West. Rather than treating these issues as given, which orthodox approaches tend to do, this book analyses the discursive relationship between US identity, foreign policy and energy security, which leads to a more nuanced and critical understanding of perceptions of China's potential threat to Western energy security.

Filling an important gap in the emerging corpus of research on energy security, this book will be particularly valuable to students and scholars of Politics, International Relations and Chinese Studies.

Andrew Stephen Campion is the Head of Research at the Atlantic Council, UK. His work focusses on the construction of security threats with an emphasis on energy and China.

Routledge contemporary China series

The Geopolitics of Red Oil

Constructing the China threat through energy security

Andrew Stephen Campion

Routledge
Taylor & Francis Group

LONDON AND NEW YORK

First published 2016
by Routledge

711 Third Avenue, New York, NY 10017
2 Park Square, Milton Park, Abingdon, Oxfordshire OX14 4RN

Routledge is an imprint of the Taylor & Francis Group, an informa business

First issued in paperback 2018

British Library Cataloguing in Publication Data
A catalogue record for this book is available from the British Library

Library of Congress Cataloging in Publication Data
Campion, Andrew Stephen.
The geopolitics of red oil : constructing the China threat through energy
security / Andrew Stephen Campion.
  pages cm. – (Routledge contemporary China series ; 140)
  Includes bibliographical references and index.
  ISBN 978-1-138-95568-4 (hardback) – ISBN 978-1-315-66616-7
  (ebook) 1. Energy security–Political aspects–China. 2. Petroleum
  industry and trade–Political aspects–China. 3. Geopolitics–China.
  4. China–Foreign relations–United States. 5. United States–Foreign
  relation–China. I. Title.
  HD9502.C62C363 2016
  333.790951–dc23   2015031286

ISBN: 978-1-138-95568-4 (hbk)
ISBN: 978-1-138-35151-6 (pbk)

Typeset in Times New Roman
by Wearset Ltd, Boldon, Tyne and Wear

To Nan and Peter, without whom I would not have started, and especially to Julie, without whom I would not have finished.

Contents

Illustrations

Figures

Tables

Acknowledgements

It is said that success has many parents, but failure is an orphan. I think it bodes well for me that I can attribute the completion of this book, and any success it may have, to a number of individuals other than myself. The book is a result of a long research project and I would like to acknowledge a few key individuals whose impact upon my work has been indelible.

At a wide level I would like to acknowledge the support of the whole research community at Newcastle University. The collegial atmosphere, which I believe is unique to the school, ensures that those of us who are lucky enough to work and do research at the institution benefit immensely from the counsel and advice of our colleagues. Specifically, I would like to acknowledge my mentors, Drs Michael Barr and Kyle Grayson. From discussions amongst colleagues I can say without any hesitation that the incredible support and guidance I received from Dr Barr and Dr Grayson has been matchless, and their scholarly influence can be heavily felt on this work.

I would also like to acknowledge Prof Christopher Dent, of the University of Leeds, and Prof Gary Rawnsley, of Aberystwyth University, who acted as my dissertation examiners and offered me encouragement to publish my work. Dr Simon Tate, of Newcastle University, was extremely supportive and helped me to prepare the manuscript through several rewrites and I owe him special thanks and appreciation for his efforts as well. I also benefited from good advice from Dr Valentina Feklyunina and Dr James Bilsland, also from Newcastle University. I am similarly grateful to Leslie Bendaly for offering her own publishing advice. Finally, I would like to thank Stephanie Rogers and Rebecca Lawrence from Routledge whose help and assistance with publishing this book has been invaluable.

Abbreviations

A2/AD	Anti-Access/Area-Denial
ANZUS	Australia, New Zealand, United States Security Treaty
APEC	Asia-Pacific Economic Cooperation
APERC	Asia Pacific Energy Research Centre
ARAMCO	Arabian American Oil Company
ARCO	Atlantic Richfield Company
ASEAN	Association of Southeast Asian Nations
BP	British Petroleum
BPD	Barrels Per Day
CCP	Chinese Communist Party
CFIUS	Committee on Foreign Investment in the United States
CNOOC	China National Offshore Oil Corporation
CNPC	China National Petroleum Corporation
CPC	Communist Party of China
CS	Copenhagen School
CSS	Critical Security Studies
CTD	China Threat Discourse
DOD	Department of Defense (US)
DOE	Department of Energy (US)
E&P	Exploration and Production
EOR	Enhanced Oil Recovery
EPA	Environmental Protection Agency (US)
ES	Energy Security
ESD	Energy Security Discourse
EU	European Union
EUR	Estimated Ultimate Recovery
FDI	Foreign Direct Investment
FTC	Federal Trade Commission (US)
G7/G8	Group of Seven/Group of Eight
GDP	Gross Domestic Product
GOP	Republican Party (Grand Old Party)
H.Res. 344	House Resolution 344
HUAC	House Un-American Activities Committee

IEA	International Energy Agency
IOC	International Oil Company
IPCC	Intergovernmental Panel on Climate Change
IR	International Relations
IRENA	International Renewable Energy Agency
ISA	International Studies Association
JDAM	Joint Direct Attack Munition
Kbd	Thousand Barrels per Day
KMT	Kuomintang
M&A	Mergers and Acquisitions
Mbd	Million Barrels per Day
MFN	Most Favoured Nation
Mtoe	Million Tonnes of Oil Equivalent
NCSL	National Conference of State Legislatures
NESA Center	Near East South Asia Center
NGL	Natural Gas Liquid
NGO	Non-Governmental Organization
NOC	National Oil Company
NSS	National Security Strategy
OCHA	United Nations Office for the Coordination of Humanitarian Affairs
OECD	Organization of Economic Co-operation and Development
OPEC	Organization of the Petroleum Exporting Countries
PLA	People's Liberation Army
PLAN	People's Liberation Army Navy
PRC	People's Republic of China
R&D	Research and Development
R/P	Reserve to Production
ROC	Republic of China
SAFE	Securing America's Future Energy
SEZ	Special Economic Zone
SIS	Secret Intelligence Service
SLOC	Sea Lane of Communication
SNL	Sandia National Laboratories
SOCAL	Standard Oil Company of California
SOS	Security of Supply
SPR	Strategic Petroleum Reserve
SUV	Sport Utility Vehicle
UN	United Nations
US	United States
USAF	United States Air Force
WMD	Weapons of Mass Destruction
WTO	World Trade Organization

1   Introduction

1.1 Where to start

Is China a threat? More specifically, is China a threat to our oil? These problem-atic and unsophisticated questions have become defining issues of international significance which have served to split China-watchers along the binary of 'China-as-threat' and 'China-as-opportunity'. Although there is little consensus about what China's rise entails, the shape it will ultimately take, and about the impact it will have, there is no dissent that it will continue to greatly affect inter-national relations and the foreign policies which constitute them. China's rise has become a prime topic of exploration and debate for Western policymakers and academics alike as they work to understand the impact of Chinese growth and even aim to decode Chinese intentions in order to calm what Robert Zoel-lick, former US Deputy Secretary of State, referred to as the "'cauldron of anxiety' over China's future".[1] Because the impact, speed, and comprehensive-ness of China's rise is unprecedented, the West has devoted great energy to studying it so as to accommodate it in the contemporary global context. The United States in particular has much invested in its desire to understand China and its rise so that it may engage with it in a manner which does not engender Sino-American conflict or jeopardize "US strategic primacy".[2] Although some look to mitigate the proliferation of China threat sentiments, these sentiments remain dominant in Western discourse.

The China threat question has become ubiquitous as it increasingly affects countless actors on the international stage in very real and tangible ways. Changes in power structures have served to entrench the question over the past two decades as China's re-emergence means different things to different actors. Nations are surveying the new and evolving multipolar and multi-centric global terrain and are adjusting to the post-unipolar climate in which US hegemony is no longer presupposed. Assuming new roles in global politics may be welcomed by those countries which benefit from increasing centrality and importance in international affairs. However, many in the West have been left reeling by the sensation that a shift is taking place where international politics are increasingly defined by Pacific rather than Atlantic issues and that the West no longer repres-ents the epicentre of geostrategic focus.

While US hegemony has become an increasingly questionable notion on a global scale, and may in fact never have reflected reality at all, the influence of the United States in international affairs remains strong and it would be folly to suggest that America's fall is nigh. Nevertheless, US power does seem to be waning in relative terms to other emerging nations and a new multipolar framework is emerging where American dominance is no longer assumed. The BRIC (Brazil, Russia, India, and China) economies, for instance, looked set to pose a credible challenge to what had become traditional US dominance in the 2000s. It was suggested that the harbinger of US, and indeed Western decline would be the twinning of the economic growth of the global south with growing influence and self-assurance of challengers and adversaries in the East. American power has, however, demonstrated its resilience and the US has maintained its place relative to pretenders including Brazil, Russia, and India whose economies have so far failed to pose the challenge which had been suggested by some. Despite US success over some of its rivals it is telling that words and phrases such as 'decline' and 'overstretch' are increasingly coupled with the United States in conversations which popularly link words and phrases such as 'rise' and 'development' to China. Thus, while its growth has cooled, global perceptions of China as a potent challenger to the US remain strong.

A recent Pew Research Center (a non-partisan think tank based in Washington, DC) report gauged public opinion from countries around the globe about perceptions of Chinese versus US power. Its findings suggest that there is a significant percentage of people around the world who are convinced that "China [will] eventually replace the U.S. as the world's leading superpower", and that these feelings are especially acute in leading Western countries.[3] Although this report did not lend a qualitative assessment to these results about whether China's growth over the US was welcome or not, the exploration of language surrounding China's rise has suggested that many in the West are wary of Chinese growth. What I explore in this book is how perceptions of Chinese growth have crystalized into a palpable China Threat Discourse, and how this has become the defining image of China in US geopolitical approaches. In particular, I explore the issue of energy security (ES), as ES has become central to nations' perceptions of their national security. As the threat from China and energy security are issues which receive frequent attention in academic and popular sources, anyone daring to add to this corpus of research must have something original and useful (and ideally interesting) to say about them. I hope you will feel that I have accomplished this in this book.

1.2 What sets this book apart

Although this book has several aims, in its broadest sense it aims to illustrate how China has been placed in an antagonistic position to the United States. While the China threat question is perhaps *the* question of our time, and while many opine about whether or not China *is* a threat, it is not a question this book will attempt to answer. The question itself is shrouded in so much ambiguity and conjecture that we must prioritize what we actually mean when we make reference to the question and speak of the China threat (CT). The book is therefore devoted to decoding China threat conventions and uncovering its discursive construction so as to illustrate how this particular discourse has affected, and continues to affect, US foreign policy.

In order to accomplish this, I examine how contemporary Sino-American relations have been both affected and represented by the relationship between certain CT perceptions and particular conventional approaches to ES which are prevalent within US discourse. These examinations will reveal how two central discourses, the China Threat Discourse (CTD) and the Energy Security Discourse (ESD), emerge from broader discussions of China and ES within everyday American conversations and debates. Having located and analysed these two central discourses, I will then examine the discursive relationship which exists between them through an analysis of the 2005 bid by the China National Offshore Oil Corporation (CNOOC) for California's Unocal Corp. A primary objective of the book is to move beyond theoretical debates to a substantive analysis of foreign policy and this case study will be used to clearly illustrate how these central discourses perform and interact with each other in a 'real world' scenario. In the decade since CNOOC made its failed bid for Unocal, there has been no comprehensive analysis of this event in the academic literature. This oversight has never been addressed in spite of the fact that the incident represents a major event which has helped to define Sino-American relations in the twenty-first century. Significant value in the book will therefore be found in the fact that it offers an in-depth exploration of an important case study which has, until now, been overlooked.

The China Threat Discourse

The importance Western actors place on studies of China and its rise is reflected by the large amount of material devoted to it and this book's bibliography is a testament to this fact. I aim to make my own contribution to the literature by almost subverting the causal approach to the subject. Rather than exploring the China threat itself, that is exploring the possibility that China's rise might unsettle Western power structures, I aim to analyse how the CT as a discourse has emerged. In this way I explore how Western power structures have actually worked to construct China as a threat rather than to explore how the CT challenges Western/US power structures. I aim to get to the bottom of the assumptions which are central to China threat arguments to demonstrate how the CT is a notion that has been actively created. In this way I do not necessarily deny or support CT arguments, but I aim to highlight that the China Threat Discourse is just that – a discourse – and that counterarguments exist to challenge it as well. In this sense, my approach is also unusual because it is unashamedly poststructuralist whereas popular approaches to case-study analysis in IR tend to remain positivist.

China threat proponents perceive a China that is capable of, and intent on, challenging Western powers. Stefan Halper, Director of the American Studies

Programme at Cambridge, effectively demonstrates how CT and ES perceptions have become entwined with one other. Halper states that the CT emerges as China "advances diplomatic, political, and economic values antithetical to those that have informed the status quo architecture" and from "its need for energy and natural resources [which] leads it to threaten its neighbors as well as regions further afield".[4] Such threat sentiments are evidenced by Emma Broomfield who claims that "The totalitarian dictatorship of the Chinese Communist Party with its expansionist goals and ruthless policies cannot co-exist in peace with the United States and its ideals of freedom and self-determination".[5] This tone is common in Western literature with writers such as Denny Roy,[6] Nancy Bernkopf Tucker,[7] David Isenberg,[8] and countless journalists and politicians engaging with CT assumptions, assumptions which are often articulated in realist terms. Employing these realist claims, John Mearsheimer states that "the US is likely to behave towards China much the way it behaved towards the Soviet Union during the Cold War".[9] Broomfield helps to further explain this phenomenon when she states: "Lest the reader forget who we are dealing with, there is an overwhelming tendency in this literature not to refer to the country as just 'China', or even the 'People's Republic of China', but repeatedly as *Communist China*".[10] Statements like this help to highlight aspects of China in which it is portrayed as not only different to the West, but also as a challenger.

Despite the fact that it provides a central theme of the book, 'is China a threat?' is not actually a question I aim to answer. Engaging with this question, in its simple and unadulterated form, requires that we be content with crass over-simplification of complex and oftentimes indefinite notions. A critical, even if cursory, look at the question exposes fundamental defects and a general opacity which is inherent to it. For instance, if we ask 'is China a threat?' do we not need to define who it is we think China threatens? While this is a seemingly logical and basic assertion, throughout the book I will demonstrate that such simple reflexivity is not common amongst China threat proponents and that the CT is instead supported by a set of unchallenged assumptions common in popular Western discourse. Returning to the question, we could also break it down further and ask what does one mean by 'threat'? Once again, I will demonstrate that while you and I may easily and habitually return to a particular notion of what we feel the idea of 'threat' entails, we are almost always guilty of leaving such assertions unarticulated. This becomes problematic when your idea of threat diverges from mine as we can no longer, by definition, be answering the same question. This disjuncture will demonstrate that threat perceptions are contextually dependent and I will explore issues of ES to locate and explore the specific CTD which has emerged as the dominant reading of China in wider US discourse.

I argue that the idea that there is a binary between China as threat and opportunity is an overly simplistic fallacy as the reality is dependent on context. Thus, while the CT tends to overwhelm competing discourses, we cannot disregard those who feel that China's re-emergence as a major power on the international stage heralds great opportunities, because these exist alongside CT perceptions in "the sweet-and-sour Sino-American relationship".[11] Nonetheless, these readings of China-as-opportunity for the West have been marginalized so that discourse of the China threat provides the fulcrum upon which perceptions of Sino-Western relations pivot and it helps to define how most foreign policy decisions about these relations are made by Western elites. Speaking from a Western point of view, even if we suggest that the theory is a fallacy, through a denial of the CT we actually acknowledge its existence which indicates that China must be, in some ways, located in opposition to the West. Even those who argue that China's rise will benefit China's neighbours and Western powers often couch their arguments in China threat language and therefore, rather than muting it, they keep the CT question alive. For instance, to promote better business relations between the UK and Britain, George Osborne, in 2013, described China "as a great opportunity, not a threat".[12] Similarly, when discussing China's economic advances, Yukon Huang, writing for the Carnegie Endowment, asks whether China's rise will be an "opportunity or threat for East Asia".[13] Loren Thompson, writing for Forbes, also gives five reasons why China will provide an opportunity for the US and will tend "to make the Middle Kingdom look less threatening tomorrow, rather than [become] a global rival for America".[14] These are but three examples among countless others, but they help to illustrate the idea that the CT as a notion has become entrenched in the collective consciousness of the West.

Although marginalized, arguments which read China as benign are just as legitimate as those which read China as a threat and I argue that both readings are part of the same conversation. However, this book examines particular arguments which twin CT perceptions with specific notions of ES in moves which further entrench perceptions of China as a threat to US interests. Highlighting this trend, Erica Downs explains how "analysts who foresee the emergence of a belligerent, revisionist state speculate that China's oil needs could prompt it to pursue destabilizing policies".[15] While particular readings of China and ES have led some, like Downs, to argue that China's rise will be destabilizing, other arguments exist which suggest the opposite and China is itself keen to promote notions of its 'peaceful development'.[16] Considering the different ways one can read China's rise it is important to demonstrate how the notion of a belligerent China has become the defining narrative in the literature which is devoted to its re-emergence. What I aim to do in this book is to explore how the idea of the CT has been constructed through discourse as well as explore what kind of politics have allowed it to thrive. So the question is not 'is China a threat?' but rather *how* has China been constructed as such.

We have to look at how the question is itself constructed because asking whether or not China is a threat presupposes that we have a clear and cohesive notion of what the CT is. Moreover, this also presupposes that we have a firm understanding of who 'we' are as well as an idea of 'who' it is that China threatens. These presuppositions are problematic as they can lead to misinterpretations of Chinese actions and even more problematically lead some into erroneous forecasting based on the belief that it is possible to capture and read Chinese intentions, or indeed the intentions of any other group or actor. Understanding the construction of the CT requires a self-reflexive examination of us – that is, the 'West'. However, as will be demonstrated in following chapters, the 'West' is itself an inherently unstable notion, but for simplicity I use it in this context in opposition to popular notions of the 'East'.[17] In order to understand the Chinese 'Other', it is essential that we understand the Western/US 'Self' from the outside in, and to examine the CTD in detail this book posits the West, and more specifically the United States, in opposition to China in an in-depth Self/Other study.[18]

Although there is general agreement amongst Western powers as to the importance of China's place in the contemporary world, there is dissent as to what it means as well as to the methods which should be used to investigate it. A significant obstacle to such investigation is the fluctuation in Western perceptions of China as the CT is the result of shifting attitudes in the West towards China. From perceptions of China as historically superior to the West,[19,20] to those of China as the 'sick man of Asia',[21] to those of a China associated with the Red Menace and the Yellow Peril,[22] to contemporary notions of Chinese development as an alternative to traditional Western development,[23] Western perceptions of China have modulated in extremes from adulation to derision. Despite this modulation, I will demonstrate that there has been a general trend of worsening opinions of China by Western actors which has occurred over the last several centuries, and which has become acute over the past four decades as the CTD has become entrenched within broader Western discourse. Because China's recent history has been so dynamic, and because much of it has been defined by revolution rather than stasis, attitudes towards it have rarely been fixed. Modern Chinese history has been shaped by a series of momentous events, including the Opium Wars of the 1800s, the 1911 revolution, the Communist victory of 1949, the Sino-Soviet Split of 1961, US retrenchment in 1972, and Deng's process of 'opening up' which started in 1978. Although these dates merely signal major events which exist amongst many other more minor ones, up to the 1980s there is a discernible trend whereby Chinese and Western interests were generally becoming more aligned with one another and China was seen, by Western powers, to be increasingly associated with the status quo. Although Chinese 'rehabilitation' continued throughout the 1980s, events in Tiananmen in 1989 served to place it in stark opposition to the West, and when we re-examine it we will see how this event represents the genesis of the CTD. In the wake of Tiananmen, the West disabused itself of ideas of Chinese normalization, and from that point on China has been viewed with suspicion – a suspicion that became increasingly coupled with anxiety about the evolution and growth of Chinese capabilities.

Though not as seismic, there were notable events which followed Tiananmen and which served to further engrain perceptions of China as a challenger to the West, with particular emphasis on the United States. The Taiwan Strait Crisis of 1996, the US bombing of the Chinese embassy in Belgrade in 1999, and the Hainan Island incident of 2001 in which a Chinese fighter collided with an American spy plane represent moments of acute distress in Sino-American relations in the decade which followed the Tiananmen crackdown. The decay in these relations enjoyed a brief period of reprieve in the aftermath of 9/11 and the ensuing American preoccupation with global terrorism. However, I argue that despite the myriad threats the US has faced in the new millennium, there is ample evidence to suggest that by 2005, China had again become seen to be the primary contender to US interests, and that the China National Offshore Oil Corporation's (CNOOC) bid for California's Unocal is indicative of this re-entrenchment. While CNOOC's bid for Unocal garnered a great deal of media attention, it has not been subject to an in-depth academic investigation, and this book helps to redress this lacuna as the event represents a pivotal moment in contemporary Sino-Western relations. In an era of consolidation in the global oil industry, the US demonstrated little reluctance to sell its assets to foreign-owned entities – that is until 2005 when CNOOC bid for Unocal. The US claimed that its energy security would be jeopardized by the sale of Unocal and the US government evoked visceral concerns about national security in order to prevent the Chinese company from acquiring Unocal. In preparation for analysis which follows, I will now give a very brief overview of the case.

In April of 2005 Chevron agreed to buy Unocal in a takeover worth $16.5 billion.[A] Both companies were American, but where Chevron was considered to be a global oil giant, despite its once formidable reputation, Unocal had become a relatively minor player. However, despite its size, Unocal possessed assets and technologies which were sought after by other companies and on 23 June CNOOC made an unsolicited bid for Unocal for $18.5 billion. Despite the higher offer made by CNOOC, the Chinese company faced significant barriers in its attempted acquisition due to the influence of the CT in American perceptions surrounding China's rapid growth in the first decade of the new century. US perceptions of an antagonistic China resulted in extensive debates as to the negative impact the sale of Unocal to CNOOC would have on US national security as arguments regarding the sale of strategic energy assets to a possible competitor country increased. Debates about the bid took place in official and non-official contexts as it was discussed at length in Congress as well as in the news media. CNOOC faced a lengthy review process by, and enough mounting pressure from the American government that it eventually withdrew its bid on 2 August, allowing Chevron its purchase of Unocal for $17.1 billion.

Through thorough examinations of the discourse surrounding it, I will argue that this bid represents a defining moment in Sino-American relations where the modern CTD became entrenched in wider American discourse. Moreover, the importance of CNOOC's bid for Unocal extends beyond the narrow scope of the case study itself as there are lessons which can be gleaned from the way the event is analysed. These lessons will be explored in the chapters which follow. With regard to the central discourses of this book, however, this event is also crucial because it represents the point at which discourse surrounding the CT became inextricably linked with that of ES. Because the emergence of the CT as the defining impression of China by the West was not abrupt, but rather a result

---

A   All figures will be in US$ unless stated otherwise.

of long discursive processes, properly understanding how it has been deployed requires placing it in context, and this book uses ES to do so.

Contextualizing the China threat: the role of energy security

Explorations of China's rise and its impact on Western foreign policy are extremely timely and can be very illuminating when done in isolation. However, such explorations are augmented by contextual analysis and I use issues of energy security to illustrate the importance of China's rise and to emphasize the role of the China Threat Discourse in US conceptions of modern China. Specifically, I look to examine how the US perceived Chinese moves for American oil to directly challenge US national interests. It is important to note that notions of energy security are no more straightforward than are those of the China threat and are, in fact, rife with definitional ambiguity. As with the CT, I will examine the nature of ES to uncover what it is and what we mean by it as well as demonstrate what the Energy Security Discourse is and how it has become the dominant reading of ES by both Western elites and audiences. Daniel Yergin, writing from a notably conventional viewpoint, states that

> energy security is driven in part by an exceedingly tight oil market and by high oil prices.... But it is also fuelled by the threat of terrorism, instability in some exporting nations, a nationalist backlash, fears of a scramble for supplies, geopolitical rivalries, and countries' fundamental need for energy to power their economic growth.[24]

Yergin's statement helps not only to demonstrate the all-encompassing nature of ES as it is inexorably linked to issues of wellbeing in the modern age, but also the importance of oil security within wider conceptions of energy security. This is especially significant as many energy analysts utilize the peak oil theory popularized by M. King Hubbert[25] to suggest that global oil supplies, necessary for our societal survival, are growing increasingly scarce.[B] Although this book addresses ES in its broadest sense, the central role of oil has resulted in frequent terminological substitution between 'energy security' and 'oil security' and the distinction should always be kept in mind.

This latter point raises an important aspect of ES in that reference in the literature is often made to a cohesive notion of 'energy security', when in fact 'energy security' is a widely contested concept.[26] Like the CT, many policymakers, analysts, and academics reference ES without articulating what they mean by it as there is a common assumption that we know what ES is and that when we invoke ideas of ES we are all talking about the same thing. This is, however, a major misconception because, as Chapter 3 will demonstrate, when we explore the term we discover there is actually little consensus as to what it is. Energy security is an expansive and contentious topic with many competing and sometimes contradictory understandings

---

B　See: Adam R. Brandt. 2007. "Testing Hubbert". *Energy Policy* 35: 3074–3088.

and narratives of what it actually refers to. The main divide within perceptions of ES results from the different roles of conventional and alternative energy resources play and the way they function in different ES discourses. Conventional ES prioritizes the security of supply (SOS) of non-renewable resources, with an emphasis on oil, and it emphasizes conceptions of safety from *in*security.[27,28] That is to say security is dependent upon the continuity of supply of energy resources, and this continuity must represent the status quo. Energy security, in this sense, is preoccupied with relative gains of resources which are limited and scarce and this places energy consumers in competition with one another.

The conflation between notions of 'energy security' and 'oil security' stem from the privileged place accorded to conventional notions of ES in Western, and especially American considerations of the term. Conversely, alternative perceptions of ES privilege renewable resources and emphasize issues including environmental and societal concerns with a focus on positive security.[29] The positive security aspect of alternative ES differs from conventional approaches because security in this sense is not gained by one actor at the expense of another, but rather security is an absolute value that can be achieved by all actors through cooperation. Thus, scarcity does not factor definitively in conceptions of alternative ES in the same way that it does for conventional ES as cooperation is thought to lead to absolute gains for all energy consumers. If, for instance, we are able to reconfigure our infrastructure to prioritize renewable sources such as wind, solar, or tidal power, competition between energy users will be greatly reduced. Therefore, perceptions of energy security are important because they can serve to bind nations together in cooperation or they can create divisions and discord, and serve to position nations in opposition to one another. These differences between negative and positive security and how they impact on perceptions of conventional and alternative ES will be explored more fully in Chapter 3.

Definitional clarity for ES is also often left wanting as those who invoke the term often misconstrue 'energy security' for 'energy policy', and this is especially true of advocates for alternative ES. It will be demonstrated that all conceptions of ES are predicated on the centrality of SOS, but some argue that peripheral concerns, including environmental and societal considerations are central to energy security when, in fact, they should usually be located under the wider umbrella of energy policy. For instance, the primary societal issue sometimes tagged with the ES label concerns the mitigation of links between political and economic underdevelopment with resource ownership, a phenomenon most commonly referred to as the 'oil curse'.[30] The institutionalization of social inequity and global inequality which stems from gaps in wealth associated with resource exploitation is something that is rightly accorded attention in academic and policymaking circles. However, in order to create clearer frameworks for analysis these studies, inextricably linked to global energy policies as they are, should be done in isolation from those which concern ES as ES is a fundamentally different thing. Definitional opacity means that energy policy can become confused with energy security when non-ES-focused considerations are considered at the expense of SOS which is, in fact, the ultimate focus and defining feature for all energy security approaches. Chapter 3 will demonstrate how the confusion between 'energy security' and 'energy policy' stems from different valuations of the constitutive elements of ES rather than differing conceptions of the term itself.

Such discrepancies help to illustrate that although there exist many conceptions of the term, those who discuss ES often assume the reader is aware of the definition they support. Despite these different understandings ES is usually deployed in its conventional guise as assumptions of the primacy of the SOS of non-renewable resources permeate debates. This has resulted in the entrenchment of conventional notions of energy security within American discourse in which oil security and SOS have become intrinsic to perceptions of US national security.[31] Oil retains a privileged position in the Western energy mix because much of our energy infrastructure has been built around it and because oil is perhaps the most strategic commodity of the day. Although oil supply security is as vital to US national security in 2015 as it was in 1973 when OPEC disrupted Western access,[32] concerns and perceptions surrounding SOS no longer tend to surround the producer-states of OPEC, but instead they stem from the growing demand from competing importer states – most notably China. As China's unprecedented growth continues, critics argue that traditional sources of energy supplies are becoming strained as they struggle to meet the growing demand, and this fear has become a defining theme of the China threat. As the US and China are both net oil importers whose ES understandings and resultant strategies are overwhelmingly conventional, these SOS concerns ensure that China and the US compete for the same scarce resources and these specific ES perceptions twin with CT perceptions to exacerbate Sino-American tensions. Therefore, the conventional notions of ES which are dominant in American discourse work to further conceptions of China as a threat to US interests because the non-renewable energy resources emphasized by the ESD are viewed as scarce and limited. When reading the situation through these optics the increasing energy demands which are inexorably linked to China's economic growth ensure that China is perceived as a competitor to the United States in a zero-sum game of energy acquisition.[33]

Perhaps the most conspicuous idea which emerges from the interaction of the China Threat and Energy Security Discourses is that China's growth will compel it to act belligerently towards its neighbours and other regional powers in order for it to secure its ever-increasing energy requirements. Of greatest significance is the argument that China's need for oil will draw it into competition with other oil importers, and particular emphasis has been placed on the possibility of conflict between China and the United States. Thus, these CT perceptions are both exacerbated by, as well as contribute to, Western ES concerns. To China threat proponents the threat has cast a pall over China's relations with its neighbours and the West, but most significantly with the United States.

If we accept that the China threat and energy security are discourses which are often deployed while rarely being defined then we are able to confront the most problematic issue which is associated with them. It is deeply troubling that reference by scholars and policymakers to notions of the CT as well as ES is often made without any suggestion or articulation as to what they mean when they invoke these terms, but this often occurs in the literature. Indeed, the very notion that the CT or ES have inherent and universally understood meanings is troublesome. In keeping, otherwise serious and thoughtful work devoted to these issues is undermined by overreliance on commonplace and often misleading expectations and beliefs. Issues surrounding the CT and ES have saturated contemporary discussions of international affairs but there is a pressing need for critical work to redress the fact that some prevalent perceptions and assumptions associated with these terms often enjoy a privileged, unchallenged, and possibly undeserved place within the literature. I therefore aim to challenge these misleading frameworks for analysis by eschewing positivist ideas of causality in order to question and explore the constructions and representations of the China threat and energy security so as to explore how the CT has become a defining motif of the twenty-first century. This examination will illustrate how the relationship between these terms has become actualized in practice, and the book will use CNOOC's failed 2005 bid for Unocal to do so.

1.3 A poststructural approach to analysis

I argue that a level of depth will be brought to the analyses of the China Threat Discourse and the Energy Security Discourse by studying them in conjunction with each other. This book utilizes an intertextual approach to illustrate how the China threat and conventional energy security perceptions expose aspects of each other which would otherwise remain unarticulated. This is not to suggest that issues surrounding China and energy can be studied alongside each other arbitrarily, but rather that there are discernible areas of overlap between them which allow for integrated analysis. I will illustrate how the CTD and the ESD work in concert to enhance China threat perceptions to US audiences. To support this claim, I examine the case study of CNOOC's bid for Unocal in 2005 as it offers a compelling illustration of the way in which conventional ES perceptions have enhanced the notion of the CT in American discourse. The ability to uncover and examine these discourses in order to explore such phenomena requires a critical approach to IR, and I will now briefly examine the central role of poststructuralism to the book's analysis.

Although a poststructuralist approach is necessarily indebted to the work of major theorists such as Foucault[C] and Derrida,[D] the approach in this book is primarily influenced by poststructuralist theorists working within the IR discipline (e.g. Ashley, Campbell, Walker). Work done by those associated with the

C See: Foucault, Michel. 1972. *The Archaeology of Knowledge and the Discourse on Language*. New York: Pantheon Books; Foucault, Michel. 2002. *Power*. London: Penguin; Foucault, Michel. 1980. *Power/Knowledge: Selected Interviews and Other Writings, 1972–1977*. Brighton: Harvester Press.

D See: Derrida, Jacques. 2001. *Writing and Difference*. London and New York: Routledge Classics.

Copenhagen School (CS) (e.g. Buzan, Waever) is also heavily referenced, with particular emphasis given to the work of Lene Hansen whose *Security as Practice* significantly influenced the theory and methodology which provides the foundation for analysis in this book.

It is useful here to clarify two points. First, I believe that poststructuralism allows us to examine the world in ways which orthodox IR approaches simply cannot. I therefore believe that poststructuralism offers us not only the best approach to IR, but the best approach to policy-relevant analysis as well. However, while I have utilized a poststructuralist foundation for analysis, I understand that many people are not so enamoured with this critical bent. Poststructuralism remains contentious amongst orthodox IR scholars and I do not seek to convert anyone to a critical cause as my aim is not to convince detractors of the value of poststructuralism. However, despite any qualms of its critics, there is a legitimate body of work in IR which is poststructuralist, and even if it is contentious, it nevertheless exists within a valid theoretical school of thought. Second, if you tend not to agree with critical approaches to IR, if you vehemently rail against them, or even if you have no prior theoretical background or associations, I feel that the primary substance of the book will retain its value for you. While poststructuralism provides the foundation upon which this study is built, it remains hidden by a policy-relevant façade. Essentially, the theory is secondary to the study, and while this has clear epistemological and ontological implications, the method of analysis does not preclude a more catholic appeal to the book's focus. The focus of the book, that is the CNOOC/Unocal case study, is one which has not been explored in depth elsewhere and I feel this exploration will be of benefit to scholars who share my predilection for critical approaches to IR as well as those who feel they are misguided.

Although this book does not aim to contribute to the theoretical debates within the IR discipline, it has been clearly informed by some of the critical work which has been done in the field. Thus, while the value of the book is primarily found in its analysis of the central discourses and the exploration of the case study, its poststructural theoretical foundations are made visible and by demonstrating the value of these foundations the book aims to promote poststructural discourse analysis as a legitimate approach to examinations of 'real world' issues. This theoretical aim is predicated on three objectives, of which the first two are clearly linked. The first objective is to demonstrate how discourse analysis is best suited to explore events such as the CNOOC/Unocal affair, and to show how Hansen's approach is best equipped to do this. The second goal is to demonstrate how a critical approach can be policy-relevant through an exploration of the interplay between the central discourses. The third objective is to demonstrate the very real impact of incremental issues to analyses of 'real world' events. I will briefly look at these in turn.

Poststructuralism and discourse analysis

First, the approach which is used in this book is indebted to the research framework Lene Hansen developed in *Security as Practice*.[34] In her book Hansen developed a poststructural approach to examine the relationship between foreign policy and identity in order to explore Western discourses of the Bosnian War. She used discourse analysis to illustrate her findings and this book will build upon her method to examine CNOOC's bid for Unocal. This method enables us to avoid the limitations inherent in orthodox approaches by engaging with discursive constructions rather than causal 'truths'. While critics argue that the poststructural emphasis on discourse results in disorganized analysis, the book will disprove this by demonstrating that a strong research framework can corral an otherwise unwieldy analysis and avoid the pitfalls of 'widening'.[E] Thus, this book will demonstrate how the research framework Hansen developed can be effectively used to examine other cases.

Hansen states that "The relationship between identity and foreign policy is at the center of poststructuralism's research agenda: foreign policies rely upon representations of identity, but it is also through the formulation of foreign policy that identities are produced and reproduced".[35] The integral role of identity to linking and differentiation and the process of Othering highlights its connection to identity and discourse. The close association of Othering to critical discourses is illustrated by Sybille De Buitrago who states,

> The other, as defined in difference to the self, can be observed in diverse contexts and dimensions within the field of international relations, as well as in other fields and in everyday life. Yet, as present as these processes of othering and self-other constructions into relations of difference and opposition are in the international arena and as essential as their scrutiny and understanding are, the analysis of such processes can only be found at the margins of past and current IR work.[36]

An exploration of Othering, and its application to the CNOOC/Unocal affair, will lend credence to the stance that it is a process best understood through poststructuralism. This, in turn, provides further evidence that poststructuralism is uniquely able to offer insight into issues of identity which positivist and constructivist approaches are unable to, and this helps to prove poststructuralsim's worth.

The Self/Other dichotomy can only be understood discursively which makes critical discourse analysis particularly effective in the exploration of the construction of the Other. Questions such as 'who we are', 'what are we in relation to others', and 'how do we relate to others' are based on a "central imaginary"[37] and this imaginary is entwined with the use of language which, is "the medium par excellence, in which these social imaginary significations become manifest and do their constitutive work".[38] Hansen also stresses the role which language plays in the construction of discourse and this is central to the intertextual models which will be explored below.

E   See: Barry Buzan, Ole Waever, and Jaap de Wilde. 1998. *Security: A New Framework for Analysis.* Boulder: Lynne Rienner Publishers, Inc.

An intriguing area of exploration in the practice of Othering has been the attempt to separate space from time, and, as will be demonstrated in Chapter 2, this occurs in China discourses. Spatial Othering is, perhaps, the most obvious way in which a Self/Other dichotomy is defined. This is due to the fact that geographical boundaries lend themselves easily to a demarcation of the Self versus the Other. Many issues do surround the political creation of geographic space and given the critical nature of this project, it would be misleading to suggest that geographical boundaries and territories comprise empirical 'truths'. However, this form of Self/Other construction routinely takes place in practice. This form of Othering, or ontopology as it was coined by Campbell on his reflections on Derrida, "refers to the articulation of being in terms of its spatial situation, the 'stable and presentable determination of a locality, the topos of a territory, native soil, city, body in general".[39] In this sense, then, it can be understood how American critics of CNOOC's bid for Unocal became alarmed when an arm of the Chinese Other effectively sought to invade the spatial locality of the US through the acquisition of Unocal.

In addition to spatial Othering, another area of critical exploration is temporal Othering.[40] As will be demonstrated in Chapter 3, this form of Othering has also been applied to China by the West which has traditionally regarded China as temporally undeveloped and backward. In its application to China, however, temporal Othering was not a self-reflexive process undertaken by the Chinese, but was a process enacted externally by the West. Daniel Vukovich explains a primary element;

> While a range of temporary – as opposed to essential – obstacles can be summoned up to explain why China is not yet free and normal, the main and seemingly most fungible one remains the Chinese Communist Party (state). Were it not for this anachronistic, evil institution, the logic goes, China would and will be becoming-the-same and joining the normal world.[41]

The 'normal world' referred to in this instance is that of the West, and Westernization, to Vukovich, is the end result of China's 'becoming sameness' with the West. But the vital point is that there is no absolute hindrance to China's normalization, only incidental characteristics, such as the party-state system, which are temporally rather than universally constituted, and which can be overcome incrementally. The importance of incremental aspects in exploring identity, foreign policy, and security, cannot be overstated as they are key to understanding the links between identity and foreign policy.[F] It is also not only positivist and constructivist approaches which lack the ability to adequately address incremental change, but also that major critical security studies (CSS) approaches, specifically securitization, fundamentally undervalue the notion of incremental change. The importance of this change will be explored below.

F See: Skonieczny, Amy. 2001. "Constructing NAFTA: Myth, Representation, and the Discursive Construction of U.S. Foreign Policy". *International Studies Quarterly* 45: 433–454.

However, returning to the ideas of Self versus Other and linking and differentiation, because positivists do not acknowledge the production of identity in this sense, they are fundamentally unable to participate in any examination of the very real discursive links between identity and foreign policy which poststructuralism is able to do. While positivist and constructivist approaches may acknowledge the impact of ideational factors on identity they perceive them as separate to material factors whereas poststructuralism sees material and ideational factors as inseparably linked as an event which is a "product of its discursive condition of emergence, something that occurs via the contestation of competing narratives".[42] A poststructural approach is essential because it challenges positivists' claims to authority over knowledge construction.[43] Poststructuralism contests orthodox IR's

> Familiar nodes of subjectivity, objectivity and conduct: they render its once seemingly evident notions of space, time and progress uncertain; and they thereby make it possible to traverse institutional limitations, expose questions and difficulties, and explore political and theoretical possibilities hitherto forgotten or deferred.[44]

One could be faulted for wondering whether such incessant self-questioning does not then render poststructuralism incoherent. However, by locating discourses through clearly defined genres and texts, Hansen developed a method based on intertextual models which serves to mitigate the chaos and disorder unfairly associated with poststructuralism.

These models are dependent on creating a framework within which texts can be organized so as to highlight the intertextual production of a discourse. David Campbell has explained that labelling theoretical approaches allows us to clearly position them in relation to each other so that we can get a clear understanding of their analytic focus. This is true with discourses as well, and using intertextual models to define the CTD and the ESD as distinct discourses allows us to approach them with a clarity we would otherwise lack. In addition, the use of models allows us to study a larger number of texts which, in turn, allows us to compare more than one discourse making for a study which is both deep and wide. By basing the study on a few clearly labelled central discourses, the pitfalls of 'widening' will be avoided. Using Hansen as a guide, this book uses four groups in the intertextual model: (1) official discourse (e.g. presidential quotations, Congressional records), (2) wider foreign policy debate (e.g. political opposition, media), (3A) cultural representations (e.g. low and high cultural artefacts), and (3B) marginal political discourses (e.g. academic output, NGO output). Organizing texts into larger groups is necessary to create a study based on central discourses which can be further defined by uncovering key texts.

These central discourses do not necessarily need to represent the popular discourse of official government policy and they do not necessarily need to be the most popular discourses. However, the CTD and the ESD do happen to be (re) articulated within both the American political establishment as well as in the

American psyche at large. By organizing texts into an intertextual research model they can be approached in an organized manner. The predominant models which will be used for the case study are models 1 and 2, which look to official discourse and wider foreign policy debate, or non-official discourse as it will be sometimes referred to.

The first model concerns documents from the Bush administration, such as the *National Security Strategy of 2002*, and quotes from officials, including the President. The second model looks at the wider political discourse surrounding US-Sino relations. Political texts, notably Congressional records and debates, articles and editorials from print media, and statements from both CNOOC and Unocal are representative texts in this model. Where models 1 and 2 will provide the texts upon which the investigation of the case study will primarily be supported, texts from models 3A and 3B will provide texts for an understanding of the wider historical context of this study and how the history of US-Sino relations also played a part in CNOOC's inability to acquire Unocal. Cultural representations are necessary to place China as an Other to the United States in a longer historical context which requires making reference to models 3A and 3B. Utilizing these models will result in an intertextual analysis which is wide, deep, and strong.

Poststructuralism and policy

The second theoretical objective of the book is to use a strong and organized poststructuralist approach to examine the interplay between the China Threat Discourse and the Energy Security Discourse in the CNOOC/Unocal case study in order to demonstrate how a critical approach can be relevant to 'real world' policy analysis. Essentially, without poststructuralism, explorations of the CTD and ESD would remain incomplete as the perceptions which are essential to foreign policy and identity, but which are often overlooked by positivist theory, would remain hidden. Essentially, popular cause-and-effect attitudes towards analysis which are often deployed in IR must be avoided in favour of analysis which embraces the discursive conversational back-and-forth which is actually more representative of the 'real world'. A discursive examination of links between China's rise and particular perceptions of ES help us to articulate how the CTD and the ESD both result from, and affect policy choices. While these central discourses may be studied in isolation, the exploration of the mutually-reinforcing relationship between them offers a much more fruitful analysis as the ESD and CTD augment each other in a very real and visible way. The examination of the case study will highlight this interplay and demonstrate how these discourses interact so that issues of ES help to amplify and codify notions of the CT in broader American discourse. Moreover, the interplay of the discourses will demonstrate how aspects of the ESD which lend themselves to empirical analysis were used by detractors of CNOOC's bid for Unocal to support the CTD and veil China threat sentiments in positivist language.

Because issues of ES lend themselves to quantitative analysis, predicated as they are on figures and numbers, it might stand to reason that an examination between ES and CT perceptions would also be quantitative. Indeed, one of the arguments I make is that ES concerns are actually used to enhance CT notions and actively legitimize the idea of the CT itself by lending objective, hard-science-based characteristics to an otherwise ethereal and subjective idea. However, rather than according validity to quantitative measurements, I will strive to demonstrate how qualitative analyses can be much more illustrative and fruitful. Thus, I use a qualitative examination of the CNOOC/Unocal case study in order to situate the larger issue of the impact the mutually-reinforcing relationship between the CTD and the ESD has on broader Sino-American relations. It will be demonstrated that other foreign takeovers of US oil companies in the years prior to 2005 succeeded where CNOOC failed and that the CNOOC/Unocal affair embodied the larger apprehension that the US felt towards China's rise in 2005. I will reveal how energy was perceived to be essential to US national interests and how China was positioned to be a contender to those interests. It will thus be argued that CNOOC was stymied in its bid because these two perceptions are incommensurable. Because other countries (e.g. Britain) were not seen to be challengers to US interests or national security, the sale of US oil companies to these foreign firms (e.g. BP) posed no security threats. However, because CNOOC is not only Chinese but also government-owned, its purchase of Unocal was anathema to US sensibilities. Frank Gaffney Jr, founder and president of the Center for Security Policy, captured threat perceptions of CNOOC's bid for Unocal when he stated,

> The purchase of Unocal by the China National Offshore Oil Corporation (CNOOC) would have adverse effects on the economic and national security interests of the United States. Such a conclusion arises from three factors: 1) The folly of abetting Communist China's effort to acquire more of the world's relatively finite energy resources. 2) The contribution this purchase would make to the PRC's efforts to dominate the vital supply of rare earth minerals. And 3) the larger and ominous Chinese strategic plan of which this purchase is emblematic.[45]

Despite the attention that was given to this bid in the American media and political debates, there has been no comprehensive analysis into *how* these assumptions surrounding the case study emerged and were mobilized. Thus, significant value in the book will be found in the manner in which it embraces intertextual discourse analysis to go beyond orthodox positivist analyses in order to redress this oversight. I will examine how the perceived geopolitical threat China poses to US interests motivated US political interference into CNOOC's bid, how this particular bid was seen to be exceptional, and importantly how CNOOC's bid for Unocal is illustrative of wider Sino-American relations.

The nature of security: incremental aspects and securitization

Finally, through my examination of the central discourses of the CT and ES I aim to highlight the importance of incremental aspects[G] in exploring identity and foreign policy. The inclusion of incremental factors to this analysis is important because while a remark or a relatively small policy implementation may seem individually insubstantial, this book will demonstrate that aggregated factors become substantive, and the central discourses are representative of this. Therefore, it is argued here that traditional, as well as some critical IR theories are too rigid and formulaic to offer the analytical insight they purport to. I will draw special attention to the work of Buzan, Waever, and de Wilde, associated with the Copenhagen School (CS), and their theory of securitization. Although traditional positivist theories which are preoccupied with big shifts in the international structure are undermined by the incremental factors which will be explored, these factors also highlight some limitations of securitization theory as the emphasis on incremental issues challenges the securitized/desecuritized binary.[H] Although many aspects of critical security studies (CSS) are highly innovative and valuable, the exploration of how the central discourses account for and are influenced by increscent and everyday Sino-American tensions demonstrates how securitization is too brittle a theory to account for the incremental factors which often coalesce to significantly influence policy choices. The examination of the case study will also demonstrate the inability for securitization theory to account for something which is positioned as a security threat within the existing rules of society as the CNOOC/Unocal affair illustrates how both China and CNOOC came to be perceived as threats without the issue venturing above the plane of 'normal' politics (i.e. a move to armed conflict). Thus, we are able to see how the CTD is primarily founded on otherwise seemingly innocuous issues such as corporate oil mergers rather than more overt threat perceptions which remain the focus of mainstream IR approaches.

While it is incredibly innovative, securitization, although it effectively describes a general trend, does not offer the structured account of the process of security it claims to. The process of securitization describes one in which referent objects move within a spectrum from having no impact, to being politicized, and finally to being securitized where the issue must be dealt with through exceptional means.[46] Although things are clearly taken to be security issues, securitization's binary between security and de-security,[47] as well as what it regards as exceptional means are concurrently (and perhaps paradoxically) too

G   Incremental factors are those which seem inconsequential in isolation (e.g. portrayal of Chinese stereotypes in media, increasing demand for Chinese oil imports) but can have an impact on policy when mobilized with other such factors within a particular discourse.

H   See: Thierry Balzacq, "The Three Faces of Securitization: Political Agency, Audience and Context". *European Journal of International Relations* 11: 171–201; Buzan, Waever, and De Wilde, *Security: A New Framework for Analysis*; Felix Ciuta, "Security and the Problem of Context: A Hermeneutical Critique of Securitisation Theory". *Review of International Studies* 35: 301–326.

definitive and too ambiguous and do not properly account for variations in the way an issue may be regarded as a security concern. In this way I argue that securitization suffers from a desire to offer an almost positivist account of security. That being said, however, I do believe that securitization's negative security approach is more convincing than the emancipatory claims to security of other critical approaches, such as the Welsh School, which value positive security.[1] As well, securitization's emphasis on the speech act is of great significance.

Buzan, Waever, and de Wilde write, "'Security' is the move that takes politics beyond the established rules of the game and frames the issue either as a special kind of politics or as above politics. Securitization can thus be seen as a more extreme version of politicization".[48] To CS sensibilities, a threat becomes extraordinary by way of a 'speech act'. The act of stating that something is a threat to one's survival helps to elevate it above the normal context of the political. Thus, there is no objective security threat as such threats are essentially political statements by those who make them as "anyone who classifies an issue as a 'security problem' makes a political rather than an analytical decision", and this highlights the importance of the discursive construction of security by instigator and audience.[49] To become fully securitized the audience must accept the securitization statement of the analyst whose role "cannot be to observe threats, but to determine how, by whom, under what circumstances, and with what consequences some issues are classified as existential threats but not others".[50] All issues and referent objects are found within a spectrum where they are either non-politicized, whereby no attention is paid to them, they are politicized, when they become part of public discourse and debate, or where they are securitized, when the issue presents an existential threat and warrants extraordinary action beyond that which can take place through standard political channels. This aspect lends itself to uncertainty as the division between politicization and securitization is blurry, and the CNOOC/Unocal case study will illustrate that something may be developed as a security threat through channels of politicization without ever having reached the level of 'securitization' in the way the authors describe it. In addition, extraordinary action is a contestable concept.

Although securitization is an enticing approach to security studies, and although some of its general tenets are plausible and dovetail well with poststructuralism, as a *theory* it is restrictive. Securitization's emphasis on the importance of speech acts, and the discursive nature between an actor who claims an issue to be a security threat and their audience who either accepts or rejects the claim, is innovative and is in keeping with poststructuralism. However, its binary between security and insecurity, conceptions of asecurity,[51] and its claim to exceptional measures and 'rules' are too rigid. Poststructuralism is, in part, so useful because it is able to absorb and explain the impact of everyday minutiae on international politics, but the strict parameters Buzan *et al.*

I   See: Booth, (1991); Wyn Jones, (2001).

define for securitization renders the process unable to account for the spectrum of issues and ways in which an issue can become positioned and accepted as a security threat beyond the theory they outline. For instance, despite the fact that it offers an exceptional example of how the US has positioned China as a security threat, securitization theory could not engage with the CNOOC/Unocal affair as the issue did not go beyond 'normal politics'. Although it could be argued that the US response broke 'norms' of the free market, it will be demonstrated that the US acted within the parameters of domestic and international law and therefore did not break any 'rules'. Thus, because it will be demonstrated that the CNOOC/Unocal affair was both a product and constitutive element of the China threat in the US, and because aspects of securitization narrow its scope of analysis to the degree it would not engage with the CNOOC/Unocal affair, the case study helps to effectively render securitization limited as a theory.

Throughout the book I will address the three objectives discussed above in order to demonstrate the value of a poststructural approach to the analysis. Rather than adopting a positivist approach to undertake a causal enquiry into the China threat and energy security I will demonstrate how discourse analysis allows us to explore much more meaningful questions. Although policymakers, journalists, and academics may devote great energies to the China threat question, I believe that a much more interesting question to ask, and indeed a much more constructive question to answer, is how has China been positioned as a threat. Rather than examining whether or not China *is* a threat to the West I will use discourse analysis to engage with *how* China has been positioned as such.

1.4 Chapter outline

Chapter 2 will begin the discourse analysis in earnest as it examines the first of two central discourses of the book. The chapter will demonstrate how the CTD has become the predominant contemporary China discourse and it will explain how it positions China as a competitor to Western/US interests. The chapter rests on analyses of perceptions of China and it will examine how the Chinese Other is contrasted to the Western, specifically US, Self. In order to understand issues which impact upon the US Self it is necessary to understand broader issues of the West as they provide the foundation for American self-perception. A juxtaposition of the West to China will then take place in order to examine the historical context of Sino-Western interaction. The focus of the analysis will then narrow further as pre-China Threat Discourses are examined. These discourses are important as they provide the basis upon which the CTD is built. The chapter will then specifically examine Sino-American relations and the role of the CTD. Constitutive elements including the Yellow Peril, the Red Menace, and China's economic and military growth will be examined to show how threat perceptions of China have been mobilized in the contemporary context. The ultimate aim of the chapter is to demonstrate how the CTD has been constructed through discourse themes which are wedded to very particular thematic and temporal readings and contexts.

Chapter 3 explores the second central discourse of the book – the ESD. Unlike the CTD which is articulated in terms of clear difference (e.g. China as threat/China as opportunity) the ESD is articulated through reference to shared measures and characteristics with other ES discourses. The differences in ES discourses emerge from their different valuations of those same sets of measures and characteristics. In order to fully explain this nuanced position the chapter will examine the contestable nature of the concept as ES is often referred to without being defined. These definitions rest on particular understandings of energy and security, and these will be examined in turn. The chapter examines the four measures which constitute all understandings of ES (i.e. resource availability, accessibility, affordability, and acceptability) and examines how different prioritizations have resulted in conventional versus alternative understandings of ES. Official and non-official accounts of ES will then be examined to see how these measures emerge in discourse and how the ESD results from a particular valuation of them. It will be demonstrated that the ESD is a particular conventional reading of ES which prioritizes short-term SOS of non-renewable resources.

Having examined the central discourses, Chapter 4 will provide the fulcrum of the book as it explores how everyday Sino-American tensions work to situate China as a challenger to US interests. The importance of the intertextual analysis will be demonstrated as the chapter explicitly addresses how the central discourses were mobilized and how they discursively functioned in order to position China as a threat to US interests with specific reference to the CNOOC/ Unocal affair. In order to illustrate the exceptional nature of the case, several successful foreign acquisitions of US oil companies which took place before CNOOC's Unocal bid are outlined. It is demonstrated that rather than being opposed to it, prior to CNOOC's bid both the American audience and elites alike were actually amenable to foreign investment into the US oil industry. The chapter will demonstrate that in the acquisition of a US oil company CNOOC failed where other foreign companies succeeded because of the particular function of the relationship between the CTD and ESD with respect to the CNOOC/ Unocal affair. The chapter will examine official and non-official US discourse to show how the CTD and ESD discursively worked to prevent CNOOC's acquisition. The central discourses will be explored in direct relation to issues of security, prosperity, and the preservation of values. This analysis of the CNOOC/ Unocal affair is important because the bid clearly demonstrates how the CT has been mobilized in American discourse, and this book provides the only in-depth analysis of the event.

The conclusion returns to the threefold contribution of the study. It will be demonstrated how Hansen's framework was effectively used to uncover the discursive nature of the central discourses of the China threat and energy security. It will also note how discourse analysis was essential to the examination of the interplay between the CTD and the ESD and how it was essential to uncover how the relationship between the central discourses is mutually-reinforcing. Importantly, the conclusion will highlight how this discursive relationship allowed us an in-depth investigation of the CNOOC/Unocal case study – a significant event which had been previously ignored. The conclusion will also reiterate the importance of poststructuralism as it was essential to the analysis which took place.

Notes

1 Jeffrey W. Legro. 2007. "What China Will Want: The Future Intentions of a Rising Power". *Perspectives on Politics*, 5(3), September 2007: 515–534, p. 515.
2 John J. Mearsheimer. 2010. "The Gathering Storm: China's Challenge to US Power in Asia". *The Chinese Journal of International Politics* 3: 381–396, p. 381.
3 Pew Research Center. 2015. *Global Indicators Database*. Retrieved 26 June 2015 from www.pewglobal.org/database/indicator/33/survey/17/.
4 Halper, Stefan. 2011. "The China Threat". *Foreign Policy*. Retrieved 8 February 2014, from www.foreignpolicy.com/articles/2011/02/22/the_china_threat, 22 February 2011.
5 Broomfield, Emma V. 2003. "Perceptions of Danger: The China Threat Theory". *Journal of Contemporary China* 12(35): 265–284, p. 265.
6 Roy, Denny. 1996. "The 'China Threat' Issue: Major Arguments". *Asian Survey* 36(8): 13.
7 Bernkopf Tucker, Nancy. 2012. *The China Threat: Memories, Myths, and Realities in the 1950s*. New York: Columbia University Press.
8 Isenberg, David. 2011. "Pentagon Talks Up China Threat". [Online] *CATO Institute*. Retrieved 8 February 2014, from www.cato.org/publications/commentary/pentagon-talks-china-threat, 30 August 2011.
9 Mearsheimer, John. 2005. "The Rise of China Will Not Be Peaceful at All". *The Australian*, 18 November.
10 Broomfield. 2003. "Perceptions of Danger: The China Threat Theory", p. 267.
11 Xia, Ming. 2006. "'China Threat' or a 'Peaceful Rise of China'?", [Online], *New York Times*, Retrieved 30 March 2015, from www.nytimes.com/ref/college/coll-china-politics-007.html.
12 "George Osborne: China 'An Opportunity, not a threat'". 2013. [Online] *BBC News*. Retrieved 27 March 2015, from www.bbc.co.uk/news/uk-politics-24512532, 13 October 2013.
13 Huang, Yukon. 2012. "China's Rise: Opportunity or Threat for East Asia?", [Online] *The Carnegie Endowment for International Peace*, Retrieved 27 March 2015, from http://carnegieendowment.org/ieb/2012/04/12/china-s-rise-opportunity-or-threat-for-east-asia, 12 April 2012.
14 Thompson, Loren. 2014. "Five Reasons China Won't Be A Big Threat To America's Global Power", [Online], *Forbes*, Retrieved 27 March 2015, from www.forbes.com/sites/lorenthompson/2014/06/06/five-reasons-china-wont-be-a-big-threat-to-americas-global-power/, 6 June 2014.
15 Downs, Erica. 2004. "The Chinese Energy Security Debate", *The China Quarterly* 177: 21–41, p. 21.
16 Xia, Ming, "'China Threat' or a 'Peaceful Rise of China'?"
17 Bonnett, Alastair. 2004. *The Idea of the West: Culture, Politics and History*. Basingstoke: Palgrave Macmillan.
18 Walker, R.B.J. 1993. *Inside/Outside: International Relations as Political Theory*. Cambridge: Cambridge University Press.
19 Trigault, Nicolas. 1942. T*he China that Was: China as Discovered by the Jesuits at the Close of the Sixteenth Century*. Milwaukee: Bruce Publishing Co.
20 Voltaire. 2000. "Splendid Secular Governance". In *Sinophiles and Sinophobes:*

Western Views of China, edited by Colin Mackerras. Oxford: Oxford University Press, 35–40.
21 Spence, Jonathan D. 1998. *The Chan's Great Continent: China in Western Minds.* New York: W.W. Norton & Company.
22 Warner, Geoffrey. 2007. "Nixon, Kissinger and the rapprochement with China, 1969–1972". *International Affairs* 83(4): 763–781.
23 Peerenboom, Randall. 2008. *China Modernizes: Threat to the West or Model for the Rest?* Oxford: Oxford University Press.
24 Yergin, Daniel. 2006. "Ensuring Energy Security". *Foreign Affairs* 85(2): 69–82, p. 69.
25 Hubbert, M. King. 1962. *Energy Resources: A Report to the Committee on Natural Resources of the National Acadamy of Sciences – National Research Council.* Washington, DC: National Acadamy of Sciences – National Research Council.
26 Winzer, Christian. 2012. "Conceptualizing Energy Security". *Energy Policy* 46: 36–48.
27 Ciuta, Felix. 2010. "Conceptual Notes on Energy Security: Total or Banal Security?" *Security Dialogue* 41(123): 21.
28 Dannreuther, Roland. 2010. "Energy Security". In *The Routledge Handbook of New Security Studies*, edited by J. Peter Burgess. London: Routledge, 144–154.
29 Brown, Stephen P.A. and Hillard G. Huntington. 2008. "Energy Security and Climate Change Protection: Complementarity or Tradeoff?" *Energy Policy* 36(9): 3510–3513.
30 Dannreuther, Roland. 2010. *International Relations Theories: Energy, Minerals and Conflict.* Polinares: EU Policy on Natural Resources, Polinares, p. 11.
31 Fuerth, Leon. 2005. "Energy, Homeland, and National Security". In *Energy and Security: Toward a New Foreign Policy Strategy*, edited by Jan. H. Kalicki and David L. Goldwyn. Baltimore: The Johns Hopkins University Press.
32 Turner, Louis. 1983. "OPEC". In *The Third Oil Shock: The Effects of Lower Oil Prices*, edited by Joan Pearce. London, Routledge & Kegan Paul Ltd.: 82–90.
33 Downs, Erica. 2004. "The Chinese Energy Security Debate". *The China Quarterly* 177: 21–41.
34 Hansen, Lene. 2006. *Security As Practice: Discourse Analysis and the Bosnian War.* New York: Routledge.
35 Ibid., p. 1.
36 De Buitrago, Sybille Reinke. 2012. *Portraying the Other in International Relations: Cases of Othering, Their Dynamics and the Potential for Transformation.* Newcastle upon Tyne: Cambridge Scholars Publishing, p. xiii.
37 Tekin, Beyza Ç. 2010. *Representations and Othering in Discourse: The Construction of Turkey in the EU Context.* Amsterdam: John Benjamins Publishing Company.
38 Gaonkar, Dilip Parameshwar. 2002. "Toward New Imaginaries: An Introduction". *Public Culture* 14(1): 1–19, p. 7.
39 Prozorov, Sergei. 2011. "The Other as Past and Present: Beyond the Logic of 'Temporal Othering' in IR Theory". *Review of International Studies* 37(3): 1273–1293, p. 1274.
40 Rumelili, Bahar. 2004. "Constructing Identity and Relating to Difference: Understanding the EU's Mode of Differentiation". *Review of International Studies* 30(1): 27–47.
41 Vukovich, Daniel F. 2012. *China and Orientalism: Western Knowledge Production and the P.R.C.* New York: Routledge, pp. 3–4.
42 Campbell, David. 2013. "Poststructuralism". In *International Relations Theories: Discipline and Diversity (3rd Edition)*, edited by Tim Dunne, Milja Kurki, and Steve Smith. New York: Oxford University Press, 223–246., p. 236.
43 Jarvis, Darryl S.L. 1999. *International Relations and the Challenge of Postmodernism: Defending the Discipline.* Columbia, SC: University of South Carolina Press, p. 90.

44 Ashley, Richard and R.B.J. Walker. 1990. "Reading Dissidence/Writing the Discip-line: Crisis and the Question of Sovereignty in International Studies". *International Studies Quarterly* 34(3): 367–416, pp. 375–376.

45 Gaffney, Frank Jr. 2005. "Gaffney Gives Hill Testemony on Unocal". [Online] *Center for Security Policy*. Retrieved 9 March 2014, from www.centerforsecuritypolicy. org/2005/07/13/gaffney-gives-hill-testimony-on-unocal-2/.

46 Buzan, Barry, Ole Waever, and Jaap de Wilde. 1998. *Security: A New Framework for Analysis*. Boulder, Colorado: Lynne Rienner Publishers, Inc., p. 21.

47 Waever, Ole. 1998. "Insecurity, Security, and Asecurity in the West European Non-War Community". In *Security Communities*, edited by Emanuel Adler and Michael Barnett. Cambridge: Cambridge University Press, 69–119, p. 69.

48 Buzan *et al.* 1998. *Security: A New Framework for Analysis*, p. 23.

49 Eriksson, Johan. 1999. "Observers or Advocates?: On the Political Role of Security Analysts". *Cooperation and Conflict* 34(3): 311–330, p. 315.

50 Ibid.

51 Waever. 1998. "Insecurity, Security, and Asecurity", p. 86.

2 Central discourses

The China Threat Discourse

2.1 Introduction

In order to understand the relationship between the China threat (CT) and energy security (ES) we must examine the discourses which surround each. This chapter will therefore study the central ones which constitute the China Threat Discourse (CTD) and the Energy Security Discourse (ESD) and it will explore how the CTD has become the prevalent discourse through which China is read in the contemporary American context. Because it allows for a deeper understanding of the relationship between the CTD and the ESD than do orthodox theories based on notions of causality, discourse analysis will be used to examine the CTD in isolation as well. This will also help to set the foundation for the explorations of discursive relationships which will take place in the chapters which follow. The CTD results from popular and often uncontested attitudes towards China, and it constitutes a central discourse of this book. To properly understand what is meant by central discourses it is useful to refer to the work of Lene Hansen who referred to 'basic discourses' in her analysis of the Bosnian War.[1] She explains that basic discourses are essentially internally coherent worldviews which exist within larger foreign policy debates and that these worldviews are central in delimiting the Self versus the Other through the relationships which exist between identity and foreign policy.[2] To avoid any misleading connotations associated with the term 'basic' I will refer to 'central discourses' in this book. In keeping with Hansen's approach I argue that understanding the CTD can best be accomplished through an exploration of identity, and by positing the Chinese Other against the US/Western Self.

Newspaper editorials illustrate how the CTD has integrated itself into the Western[A] popular consciousness. For instance, a *Washington Post* editorial written during a time of escalating tensions between China and Taiwan laments US "strategic ambiguity" and "public accession" to the Chinese "dictatorship['s] … bullying posture".[3] More recently William Wan has claimed that PLA double-digit

---

A The 'West' is a broad term and is discussed in section 3.2. This book refers to Western European states (especially those directly linked to the Enlightenment project) as well as North American states as 'the West'.

growth has resulted in US readings of the Chinese military as "a rapidly modernizing and expanding force that could one day rival, or even worse, overtake that of the United States".[4] While the military aspect of the CTD is highly visible, the economic aspect of the CTD is also potent, so much so that Robert Samuelson, columnist for the *Washington Post*, even argues that China might actually destabilize the world economy.[5] While such empirically measurable elements contribute to the CTD, other ideational aspects also play a significant role. According to Joshua Cooper Ramo, a China specialist and former senior editor of *Time* magazine, the greatest threat China poses to the West, and the US in particular, is its national image. The essential problem is that while China may regard itself as a peaceful, inward-looking, and benign regional power, its neighbours may perceive it to be a belligerent and ambitiously expansionist power. It is significantly problematic that wildly varying perceptions of China by itself and its anxious neighbours may be reached through different readings of the same phenomena. Thus, there is a great gulf which exists between Western perceptions of China and the Chinese self-image, and this book is preoccupied with the examination of how Western and US perceptions came to populate the critical side of this gulf.

However, the fallacy of equating popularity with accuracy must be avoided as the prevalence of the CTD does not necessarily make it 'true'. Some actors may be right to fear China but we tend to take these fears at face value without examining what constitutes them in the first place.[6] Such superficial readings can be dangerous and misleading. For instance, perceptions of China as a revisionist and expansionist power are challenged by those, such as Professor Jin Canrong of Renmin University who states that "China remains an inward-looking country. It is essentially not interested in the outside world, except to make money. So the West should not expect too much from China in global affairs".[7] When past events, such as the Taiwan Strait Crisis in 1996, are examined in conjunction with more recent events, such as the Sino-Japanese dispute over the Diaoyutai/ Senkaku Islands,[8] they could constitute evidence to support perceptions of China as expansionist. Ted Carpenter and James Dorn, writing for the CATO Institute, demonstrate the way events and issues can become conflated when they argue that Taiwan remains a flashpoint and that any move by the Taiwanese towards independence will cause China to respond militarily.[9] However, we should not lose sight of the fact that while China may feel entitled to react militarily in such a situation it may not feel compelled to do so.[10]

Despite overlapping territorial claims, the CCP has generally been content to live within its own borders and it is keen to promote this image. However, with growing tensions surrounding Chinese territorial claims in the South China Sea the promotion of a passive image of itself by China to its neighbours and other regional powers is becoming increasingly difficult. Nonetheless, the People's Liberation Army Navy (PLAN) has taken steps to ensure neighbouring powers that China is "non-global, non-expansionist, and non-pre-emptive" and that its naval interests are only regional.[11] Despite such measures, however, other powers do tend to remain wary of Chinese intentions as the CTD has become the primary conduit through which perceptions of China are shaped and expressed.

Discourse analysis will allow us to examine how certain assumptions central to the CTD have been accepted.

2.1.1 *Approaching the China Threat Discourse*

The China Threat Discourse must be situated within the larger context of discourses on China to be understood. In order to provide this context, we will start with the most general enquiries surrounding relations between China and the West in their broadest sense. This will allow us to concentrate on specific US perceptions of China as a threat afterwards. Narrowing the focus of analysis in this way will help to locate the CTD amongst other competing, yet ultimately marginalized, China discourses. Understanding the CTD requires properly understanding its spatial and temporal aspects and origins which is to say we need to locate the time and location of the earlier discourses from which the CTD emerged. Being able to extrapolate the China threat from issues of time and space will allow us to look at the ways in which Western, and in later periods, specifically American, perceptions of China have been formed.

To understand US perceptions of a China threat a logical starting point would be to analyse a fundamental aspect of its spatial location and to examine how the notion of 'the West' will be utilized in this book. This examination is important because without a definition of the term, it would remain ambiguous and contentious and reference to 'the West' would undermine the book's analysis.

A chronology of China discourses will then be outlined as the CTD is the product of several discourses which have temporally preceded it, and although they share similarities, there are significant areas of departure which allow the CTD to stand apart from previous discourses. However, properly understanding how the CTD has emerged, and how it is differentiated from other Western discourses of China will not be done temporally, but rather thematically. Through a chronological examination, thematic elements will emerge which are constant throughout these discourses while others will disappear.[B] This is due to the fact that Western interests and goals in China have changed and evolved over the centuries the West has been involved with it.

The examination will begin with initial sustained European interaction with the Chinese in the sixteenth and seventeenth centuries which will be explored in section 2.3. Although European contact with China was made centuries earlier, the constant presence of Europeans in China from the sixteenth century onwards is vital. Despite the fact that this book scrutinizes Sino-American relations, a focus on these early European contacts is important as they represent some of the foundational Western interaction which occurred with China before the founding of the United States itself. The examination will then turn to the eighteenth and nineteenth centuries which witnessed greater Western exploration into China and a change in perceptions by many Europeans and Americans of China.

B A similar approach to textual analysis in the ESD is used in Chapter 3.

The twentieth century will then be studied in section 2.4 as it represented a time of tumultuous change both within and without of China. Understanding these periods will help to understand perceptions of China's global role in the twenty-first century in which the CTD has become the primary Western discourse of China. Section 2.5 will be devoted to an in-depth analysis of the CTD.

2.2 The 'West' and China

Although this book analyses the Sino-American relationship specifically, doing so requires an understanding of the relationships that existed between China and the West (i.e. Europe) before the United States even came into being. The fundamental and formative perceptions of China borne of this time are essential to understanding specifically American perceptions which followed. Because Sino-American relations rest upon the foundations of broader Western (European) perceptions of China, these latter insights, and their origins, must be thoroughly assessed in order to account for the former.

Until the formal creation of the United States in the late 1700s, Western relations with China were solely represented by Europeans whose contact with China had begun centuries earlier. Although it is problematic to refer to the West as any unified body of shared tenets, goals, or peoples as it is an unfixed and variable notion, in a Saidian sense the West can be generally used in contrast with 'Oriental' China as the Orient and the Western Occident have been instrumental in constituting each other through their contrasting imageries and characteristics. Because contemporary Sino-American relations and popular perceptions and representations of China find their genesis in these first European contacts with China, analyses of the China threat vis-à-vis the US must begin with a broader spatial and temporal understanding of Sino-*Western* relations. Although the US and the West may be analysed separately, it is important to emphasize that the United States ultimately falls under the umbrella of 'the West'.

Western perceptions of China have varied wildly since initial European contact with China and it is essential to highlight the extremities of the perceptions of China by the West. Regarding America, the historian A.T. Steele states feelings towards China are often in flux and "The ups and downs of public opinion on China becomes understandable only against the historical background and the heritage of assumptions, expectations, emotions, traditions and even illusions and legends which have contributed to our present attitudes".[12] At its widest, the historical background portrays not so much an oscillation between admiring and disparaging accounts of China by the West but rather an established, measured, and clearly-defined shift from Western admiration, especially pronounced by the Jesuit writers in the seventeenth century, to derision of China which became commonplace in the nineteenth and twentieth centuries. Examining how notions of a threatening China emerged in the United States from earlier European understandings of China as a peaceful and sophisticated society is central to this book.

2.2.1   *Historical context of Sino-Western interaction*

Competing and contrasting notions have always existed in the Western mind-set about China. 'Barbaric' China has clashed with 'civilized' China, and 'backwards' has contended with 'developed' China.[13] Never at one time has there been a singular voice representative of all Western perceptions of China, and despite the prevalence of the CTD, this is equally true now. However, different periods do seem to favour one extreme over the other, eliminating an entirely uniform notion of China. Broadly speaking, however, the image of China as seen through Western optics has been one of steady corrosion. Western accounts of China from the fifteenth and sixteenth centuries are notable for the admiration and respect the authors generally bestowed upon Chinese society. Reading texts by European explorers and missionaries of the time, such as Marco Polo and Mateo Ricci, gives one the impression of near wonderment with which they viewed alien aspects of China. Western accounts of China began to change during the Enlightenment and significant criticisms began to emerge as China's perceived superiority was undermined. Finally, readings of Western accounts of China from the twentieth and twenty-first centuries often demonstrate the extreme degree to which Western perceptions of China have soured. These texts, often written with a distinctly American voice, unabashedly portray China as an antagonist to Western/US interests. The substitution of admiration for derision and fear of China has been a gradual process which has progressed, very incrementally, over a number of centuries. In order to properly engage with Western and US readings of China it is more important to engage with thematic rather than temporal elements of the China threat. Prominent thematic readings of China are explored in the next section.

2.3   Pre-China Threat Discourses

Properly examining Sino-Western themes requires that we transcend a chronology as thematic analysis has a much more powerful impact. The China discourses that will be referred to are the Respected China Discourse, the Disrespected China Discourses (including Chinese-as-barbaric, Chinese-as-feminized, and Chinese-as-infantilized themes), and the China Threat Discourse itself, the latter of which will be examined in greater detail in section 2.4. Perceptions are intrinsically linked to identity construction because self-perception and perceptions of Others create the limits of the Self which is essential to the creation of political entities, boundaries, and space. The China discourses referred to will illuminate as much about the US Self as they do of the Chinese Other, placing identity construction at the centre of analysis. Each discourse will have spatial, temporal, and ethical imperatives which will be explored in each in order to expand upon themes of linking and differentiation through which the West/US has read, and continues to read China.

2.3.1 *The Respected China Discourse*

The first period of Sino-Western interaction was largely defined by European admiration for China. John Fairbank explains that "China's first contact with Europe was … extraordinarily fruitful", and this was true for both China and the West.[14] The Respected China Discourse places great spatial, temporal, and ethical distance between the Chinese Other and Western Self. Although Chinese and Western entities are never static, I argue that the centuries-long, ever-changing nature between them is more the result of changes in the West than of those within China, with specific emphasis on epistemic shifts in Western knowledge construction. Different forms of knowledge offer different views of China, and early Western knowledge-power structures lent themselves to complimentary readings of China.

Early European representations of China are not especially deep, partly because there are many fewer Western accounts of China from this period than later periods. Because of this, each text has a disproportionately large impact upon readers than do later texts which help to form other discourses. As few Westerners had contact with China prior to the sixteenth century, perceptions of the country were largely influenced by the exoticism with which early explorers viewed it. The accounts of Marco Polo, whose opinion of China was overwhelmingly positive, are indicative of this. Speaking of the grandeur of Hangzhou, the city he calls Kin-sai, Polo claims visitors become "intoxicated with sensual pleasures, [and] when they return to their homes they report that they have been in Kin-sai, or the celestial city, and pant for the time when they may be enabled to revisit paradise".[15] Polo's wonderment of China extended beyond the great walls and canals of Hangzhou, a city at least 20 times the size of his native Venice, but his impression of urban life in China left him most impressed.[16] Polo's *The Description of the World* "was the first such work by a Westerner to claim to look at China from the inside, and the force of the narrative description was strong enough to imprint itself in Western minds down to our time".[17] The immense influence of Polo's account, and his unique perspective, helps to reinforce the exotic veneer of Imperial China, although the difference between reading China and reading Polo becomes somewhat blurred. While particular claims in *The Description of the World* may not be replicated in other texts, other accounts of early Western interactions with China reinforce the respect which Polo showed the Chinese. These similar impressions help to construct cohesion in the Respected China Discourse.

The nature of early Western accounts is also essential to understanding the Respected China Discourse. During the earliest Sino-Western interactions the West was more representative of Christendom and universal monarchy than that of a modern state-system upon which contemporary notions of the West are based. The modern Western nation-state would emerge alongside the rationalist revolution in later centuries. Although European, the earliest Western accounts were largely those of missionaries and explorers whose primary allegiances were to Christian hierarchies, and whose ideals underscored their interactions with the

Chinese. Although the patrons of explorations and missions were often European elites, they were themselves not free of influence from the Church. I argue that the prevalence of religious motivation helps to explain how China was read and why it was accorded respect in this discourse.

Whereas contemporary issues surrounding Sino-Western relations largely relate to materialistic aims (e.g. economic growth, military security) the interest of missionaries in China was largely focused on the ideological goal of spreading the Christian gospel. Although trade was always an intrinsic aspect of Western relations with China, much of the early interaction was centred on religious propagation as European Christians aimed to convert the Chinese infidel. This goal is evidenced by the desire of St. Francis Xavier to travel to China in 1552 to show that through "the labours of our Society, the Chinese and Japanese will abandon their idolatrous superstitions and adore Jesus Christ, the Saviour of all nations".[18] Even non-missionaries, such as Magellan, devoted energies to the Christian conversion of Southeast Asians.[19] These Westerners were not primarily focused on the material elements of China which would be central to later, more derogatory discourses on China.

The Respected China Discourse positions China in a positive light as Western accounts explain how Westerners of the time were impressed with many aspects of Chinese culture and the ways in which it was copasetic with many Christian and Western sensibilities. The experience with China was different than that of other targets of Christian evangelism in the sense that China was not seen as a 'barbaric' country. Chinese society was well ordered and although the Chinese were heathens, the advanced level of Chinese culture was not lost on the West-erners. For instance, Matteo Ricci, an early Jesuit missionary, was impressed by the fact that the Chinese, like the Europeans, used chairs and tables to work and dine.[20] Although a seemingly insignificant detail, it does serve to highlight that there were points of similarity between East and West and that the Chinese Other was not as Othered to the West as were other societies, such as those in the Americas. Although different to be sure, Chinese culture was seen to be impres-sive by Western standards as can be evidenced by flattering Western accounts of Chinese intelligence, Chinese desire for peace and freedom, and Chinese governance.

Intelligence in the Respected China Discourse

The Respected China Discourse emerges, in part, from texts which show clear admiration for Chinese intelligence and ingenuity. This can be witnessed in the accounts of St. Francis Xavier who states that China was "a most opulent empire, abounding in everything necessary for human life.... Its people are remarkable for intelligence, and employ themselves in study".[21] Alessandro Valignano, a Jesuit missionary, also expressed admiration for the learned Chinese culture and stated that "The Chinese cultivate letters seriously and hold learning in high regard.... The Chinese are alert, enterprising, and lively in their actions".[22] The observations of Matteo Ricci underscored the importance the

Chinese placed on education which Valignano also witnessed. Despite a plethora of local dialects which were impenetrable by way of the spoken word alone, the characters which comprised the Chinese written word were universal. Writing was so widespread that its influence is said to have surpassed that of Latin in Europe, and "The Chinese would often use their fingers to draw the characters corresponding to their spoken words in the air or on the palm of their hand, thus showing that it was the written rather than the spoken language that unified the empire".[23] Written literature was also not just a novelty for the rich and well educated, but was common among the Chinese people as a whole, something which was inconceivable in the West.[24] Ricci's view of the Chinese as a literate people was confirmed by the amount of literature available to all Chinese including newspapers and gazettes.[25] Whereas books were a precious commodity in Europe, they penetrated all levels of Chinese society which was extremely literate by Western standards.

Peacefulness and freedom in the Respected China Discourse

Western admiration for the peaceful and accepting nature of China provides another pillar of the Respected China Discourse. St. Francis Xavier provides another illustrative account of the respected attitude towards the Chinese by stating "They have generally kind open dispositions, and are lovers of peace, which flourishes and is firmly established among them, without any fears of wars".[26] Juan Gonzalez de Mendoza provides another widely read volume consisting of different accounts of China by various European travellers at the time. Mendoza praises Chinese society for its distain of idleness and its admiration for industriousness, a quality which was seen to be common among Chinese citizenry.[27] Despite its industriousness Mendoza also admired China for its restraint and that it did not overreach itself, and was, instead, very inward looking. Mendoza explains this Chinese self-discipline and moderation when he states that "The king doth content himself only with his own kingdom (as one that is held the wisest in all the world)".[28] As such, China was perceived as a peaceful country which was free of war; something that had defined much of Europe's history and something that still defined relations between European powers. A major Jesuit contributor to early Western perceptions of China was Jean Babtiste Du Halde who illustrated stark contrasts between Europe and China to highlight what he saw as the relative backwardness of European ways. Du Halde provided one of the most robust complements of the Chinese character when he wrote that

> The *Chinese* in general are mild, tractable, and humane; there is a great deal of Affability in their Air and Manner, and nothing rough, harsh, or passionate ... they learn betimes to become Masters of themselves, and value themselves in being more polite and more civiliz'd than other Nations.[29]

Christian accounts of China were also complementary of the religious freedom which was accorded to the Chinese and Europeans alike by Chinese elites. Early Jesuits and other missionaries were not impeded from worshiping and even promoting the Christian gospel amongst the Chinese as Chinese elites did not regard religious worship as being detrimental to the functioning of imperial society. This religious freedom later helped to inform central aspects of the Enlightenment.[C]

Governance in the Respected China Discourse

For all the points of their admiration, it is conceivable that the early Western visitors to China were most impressed with its advanced system of governance which arguably provided the basis of the modern state-systems of Europe. An innovative feature of Chinese government was its examination system. Rather than entrusting public office to highborn classes as was the custom in Europe, the Chinese valued aptitude over breeding and entrusted public office to those best able to serve as proven through a process of examination.[30] This was a revolutionary idea to the Europeans and one which they would adopt to good effect over the coming centuries, although the Europeans' model would place great importance on scientific learning, which was not traditionally a strong point of the Chinese system and would eventually stymie China's influence in international affairs. Although borders would be changed and territorial disputes would arise, largely speaking it seemed that through the bureaucracy the system engendered, Imperial China was content to live within its own boundaries, at the centre of a system of tributaries which remained committed to China because of the trickle-down benefits of its power and inertia. China's diplomatic power rested on a mix of economy and security to promote a perception of itself as a civilizing leader for its region[31]; a self-perception common to all dominant empires at their height of power. Reading China through early European texts it seems that such efficient governance allowed a truly civilized China to flourish amid dependent powers. Returning to Marco Polo, he

> was extremely positive about China, writing in glowing terms of its governance and cities [and] he describes China as a land of great prosperity and flourishing commerce and claims that the emperor took a personal interest in the well-being of his subjects.[32]

Such glowing reports of China would become increasingly scarce in the centuries which followed.

2.3.2 Shifting perceptions of China

There was no abrupt or definitive event which signalled the shift in Western perceptions of China from adulation to contempt, but a gradual shift did take place

---

C It is interesting to note how this religious freedom evaporated as the CTD became more entrenched.

in Europe during the seventeenth and eighteenth centuries. This particular period saw a revolution in Western thinking, the impact of which cannot be overstated. The seventeenth and eighteenth centuries witnessed Europe emerge from the Dark Ages as the Enlightenment fundamentally altered the epistemic knowledge-construction in the West to prioritize reason and scientific method. Sebastian Conrad describes the dominant view when he states "The Enlightenment appears as an original and autonomous product of Europe, deeply embedded in [its] cultural traditions".[33] The Enlightenment not only represented an overall acceleration in the accumulation of knowledge in Europe, but it also represented a shift in the *ways* in which knowledge itself was understood. There was a revolution in epistemological enquiry in the West, the likes of which did not take place in China.

The Respected China Discourse reads China as an innovative and developed country which was traditionally more advanced than the West. However, during the Enlightenment Europe's advances caused China to retrograde in relative terms and it became perceived as static. Fairbank and Goldman use the abacus as an example of how earlier innovation may have ultimately hindered the development of China. They argue that while the abacus is an ingenious device, as its mathematical use is limited to simple arithmetic its widespread use amongst the Chinese ensured that they forewent the more complex mathematical innovations which were taking place in the West. Thus, China may actually have suffered long-term development because of short-term innovation. The argument is not that it was beyond the ability of the Chinese to understand complex thought. Rather, Western-style science was simply not considered to be important to Chinese education which favoured philosophy and literature, which provided the building blocks of government education, the crowning achievement of Chinese civilization.[34] Europe differed because "Europe had inherited ways of thought that made it more ready for scientific thinking",[35] and even "medieval scholastic philosophers demonstrated remarkable flexibility in reconciling revealed religion with natural-law-based science – something other societies found difficult".[36] Thus, although popular perceptions of medieval Europe are of scholastic 'darkness', Europe did have a scientific tradition which exploded as Chinese innovation declined and "the very superiority achieved by Song China would become by 1800 a source of her backwardness" as earlier innovation engendered innovative complacency.[37] Thus, the outbreak of new thinking in Europe accelerated its development and helped the West to leapfrog China.

Modern Europe developed as Christendom declined. Reason, in modern Europe, replaced religion and tradition and attitudes towards China began to change as the West read it through a new epistemic lens. The West began to criticize China for its lack of progress, and if power is associated with knowledge, the Enlightenment represents the time in which the West began to cement its power over China. Despite the fact that the Enlightenment heralded Western dominance over China, it must be noted that the Enlightenment owed a debt of gratitude to Chinese thought as well. The changing perceptions of China during the Enlightenment were gradual and China was, at times, simultaneously revered and reviled by different factions in the West.

Voltaire "was undoubtedly the most influential of the French philosophers and historians of his day and his role in creating a positive image of China in the eighteenth century was of the utmost significance".[38] Unlike Europe where divisions between political and religious rule were often indistinct, Voltaire revelled in the Confucian approach to secular governance in China. The separation of church and state would become a major point of contention in the political revolution which would upset powers in France, Britain, and America, and this division of powers was inspired by the Chinese Imperial courts. Voltaire was enthusiastic about this separation of powers and China's success at delimiting the two caused Voltaire to proclaim that "Here the Chinese are particularly superior to all the nations of the universe".[39]

Changing perceptions of China also meant that Chinese critics became more numerous and vocal. Even texts from pro-China thinkers tend to display a growing condescension during this period. For instance, Voltaire writes that,

> It is sufficiently known that [the Chinese] are, at the present day, what we all were three hundred years ago, very ignorant reasoners. The most learned Chinese is like one of the learned of Europe in the fifteenth century, in possession of his Aristotle. But it is possible to be a very bad natural philosopher, and at the same time an excellent moralist. It is, in fact, in morality ... in the necessary arts of life, that the Chinese have made such advances towards perfection.[40]

While he is complementary of certain Chinese characteristics, Voltaire's tone of language suggests an assumed Western superiority over China.

The Hakluyt Society's 1853 translation of Mendoza's work edited by Sir George T. Staunton is similarly patronizing. Mendoza's *History* compiles the

> early glimmerings of information which Europe obtained respecting a country so removed from the civilized world, by its geographical position and ethnological peculiarities, as China, yet so marvellously in advance of it at the times of which we speak, both in its intellectual and moral developments.[41]

This quote by Staunton is very interesting in the way presents China to the reader as it reflects the often contradictory opinions of Europeans towards China. Mendoza's work, completed in 1585, was very much a product of its time in that it saw China as both alien and awesome. Staunton's, however, is equally a work of its time and the introduction reflects the growing disdain with which Europe viewed China by 1853. Staunton does not disregard the praise paid to Chinese society in Mendoza's work as he admits that China was "marvellously in advance of [Europe] at the times of which we speak", but it is notable that when he writes in 1853 he stresses China's separation from the 'civilized' world. In delineating a divide between China and the civilized West he clearly alludes to the role reversal of China and the West (in Western minds) which had occurred by the mid-nineteenth century.

Adam Smith provides an anti-China counterpoint to Voltaire's pro-China stance. Smith's greatest contribution to the Enlightenment was his innovation in economic thought. It is telling that although he regarded China as naturally rich and fertile, his appreciation for the country was phrased in such a way as to suggest he equated its worth with the monetary value it represented, and his great frustration with China was based on the notion that it did not exploit its economic potential. When one reads Smith, one feels a measure of resentment directed towards China. Smith writes that "China has been long one of the richest, that is, one of the most fertile, best cultivated, most industrious, and most populous countries in the world. It seems, however, to have been long stationary".[42] He states that Marco Polo wrote of the same natural wealth, but Smith also notes that China had done nothing to take advantage of those riches in the intervening centuries. He also levels criticism against China for its squalor and poverty and argues that the Chinese only work hard as their impoverishment demands it of them. He writes that

> The account of all travellers, inconsistent in many other respects, agree in the low wages of labour, and in the difficulty which a labourer finds in bringing up a family in China. If by digging the ground a whole day he can get what will purchase a small quantity of rice in the evening, he is contented.[43]

Far from painting an idealized picture of China, such a statement suggests the desperate underdevelopment of China.

Western criticism of China also grew along with the growing trade imbalance between them. Smith illustrates this frustration: "The Chinese have little respect for foreign trade. Your beggarly commerce! Was the language in which the mandarins of Pekin used … concerning it".[44] Although a voracious appetite existed for Chinese goods in the West, the Chinese market did not reciprocate until the British East India's introduction of opium.

2.3.3 Disrespected China Discourses

As Western development continued and industrialization began in earnest, perceptions of China continued to decay. New discourses emerged which portrayed China as a weak and undeveloped power which was temporally distant to the West. The spatial distance between China and the West did, however, continue to diminish as greater numbers of Westerners, including Americans, travelled to China. Thus, greater numbers of texts emerged and the variance and detail of the accounts increased. China, as read through the Disrespected China Discourses, was not feared by the West, but rather pitied. There are three clear themes as to how China was popularly read which emerged in the Disrespected China Discourse wherein it was infantilized, barbarized, and feminized.

Infantilized and Barbaric China Discourses

Lord Macartney, British ambassador to the Chinese court, summarized Chinese-as-barbarous sentiment when he stated that "A nation that does not advance must retrograde, and finally fall back to barbarism and misery".[45] Macartney perceived the Chinese to be a semi-barbarous people whose reluctance or inability to grasp the scientific and technological underpinnings of industrialization ensured their subservience to the West. Chinese customs which had once charmed Westerners were now often viewed negatively and the plight of Chinese women, infanticide, and the despotic nature of Chinese government became themes associated with perceptions of Chinese backwardness.[46] The practice of female infanticide was closely related to the poor station in Chinese society women were seen to occupy. Justus Doolittle, an American missionary to China, stated "No doubt infanticide is more common in some localities and provinces than in others. But ... it is tolerated by the government".[47] Although perceptions of a morally elevated China had by this time been largely extinguished, female infanticide was seen to be a practice born of barbaric necessity as it was thought that poverty in China precluded families from the luxury of having girls. Doolittle goes on to state that,

> In China the doctrine of filial piety is highly lauded, and children of both sexes are required by law and by the usages of society to render the most implicit and even abject deference to the will of their parents. But parents are permitted to discriminate between the sex of their helpless offspring, destroying the female *ad libitum*, and lavishing on the male their care and love. How singularly and emphatically are they 'without natural affection' as regards this subject![48]

Not only did Western perceptions of the strong moral and ethical fibre of China change, but Chinese government became seen as corrupt and ineffective. Charles Beresford visited China to assess the political and social environment and to report on the health and future of British interests there. He suggests that there was great anxiety and "this existing sense of insecurity is due to the effete condition of the Chinese Government, its corruption, and poverty".[49] The West's responsibility to China thus became one of a civilizing mission.

Rodney Gilbert helped to promote the perception of Chinese-as-infantile and stated

> The difference between the Eastern and Western mentality is precisely the same as the difference between the puerile and adult mind ... most of China's ills have grown out of her own and our failure to appreciate that the Chinese mind is a child's mind – the mind of a precocious child at its best and worst.[50]

Gilbert continues,

> there are nations that cannot government themselves, but must have a master, just as there are men in every community that need a guardian and

are a menace to the community if granted the unqualified 'right to life, liberty, and the pursuit of happiness'.[51]

This infantilized China discourse suggests that the West had a responsibility to govern China and provide a parental and police presence as the Chinese were unable to tend to their own affairs due to an incapacity for change.

In an account of Baron George Anson's visit to Guangdong province, Richard Walter and Benjamin Robins offer a view of the Chinese far removed from the moral and ethically superior peoples as described by earlier Sinophiles.[52] Walter and Robins "frequently scoff at the prevailing view of China in works such as those by Voltaire and the Jesuits".[53] They state,

> That the Chinese are a very ingenious and industrious people is sufficiently evinced from the great number of curious manufactures which are established amongst them, and which are eagerly sought for by the most distant nations; but though skill in the handicraft arts seems to be the most important qualification of this people ... they are in numerous instances incapable of rivalling the mechanic dexterity of the Europeans. Indeed, their principal excellency seems to be imitation.[54]

Such views contrasted greatly with earlier notions of the Chinese as inventive rather than imitative. China was responsible for the invention of many products whose usurpation by Western industry would fundamentally, without exaggeration, change the world. Ironically, Chinese inventions, notably gunpowder, would subjugate it to the West as Western powers appropriated them.[55] Gunpowder led to the gunboat diplomacy which would place China under the thumb of the West. Rather than being notable as a strong industrial power, China became to be seen as a master of handicrafts and ornamental commodities whose value to the West was purely aesthetic. Although seemingly innocuous commodities, the procurement and use of Chinese porcelain and tea by Western powers led to a bonanza which not only transformed the Western powers, but further transformed perceptions of China as well.

Feminized China Discourse

The feminization of China is best illustrated by the Western desire for Chinese tea and porcelain. As Western power grew, so did its *lust* for Chinese commodities and power over China itself. The more Chinese commodities, most notably tea and porcelain, inundated Western markets in the eighteenth and nineteenth centuries, the more the West was seduced by China. This occurred in the United States to such a great degree that Chinese imports became to be seen as a perilous threat to US self-determination as these commodities became highly valued and became essential to trade with China.[56] Although eighteenth century America was not yet an independent entity, it wished to show its presence as an international player by involving itself in international trade. Caroline Frank states that

American consumers, like Europeans, in the seventeenth century wanted access to Asian luxury goods and wealth, and they developed and legitimated their own methods and terminology for doing so specific to their perceived role and place within an Atlantic political economy.[57]

The desire to assert its independence placed America at odds with Britain as Britain's East India Company monopolized the trade of Chinese commodities. Wary of growing independence, Britain labelled American traders as pirates. Frank states that "ultimately, Americans participated in the China craze in a way that was distinctive both to their place in the British Empire *and* in the world".[58]

As their place in the British Empire became less palatable to Americans, so too did their dependence upon the import of Chinese tea and porcelain, which was strictly regulated by the British. Although Britain had repealed taxes on most imports to the US due to pressure from those who argued for no taxation without representation, taxes remained levied on tea imports.[59] This led to the dumping of British tea into Boston Harbour by American rebels in 1773, and "From that moment on, it became a patriotic American patriotic duty to avoid tea" with the Continental Congress even passing a resolution against its consumption.[60] The 'taking' of tea soon became synonymous with idleness and femininity. John Adams wrote to his wife stating "Tea must be universally renounced ... and I must be weaned, the sooner the better".[61] Both China and china became seen to be fragile and effeminate and subject to an existential contradiction in that one acquired a measure of masculinity through ownership, yet the items themselves became synonymous with femininity and epitomized female delicacy. Thus, while great industry was once spent attaining these items, "gradually the masculine aura once granted china in England gave way as porcelain vases and tea sets ... became the defining trope of a languid, irresponsible femininity".[62] Subsequently, porcelain and tea became associated with female pastimes in Western culture. China became feminized in American discourses in order to make Chinese commodities less appealing, thus lessening American dependence on Britain. It is also interesting to note that the Western term for the Chinese-inspired decorative aesthetic popular at the time was 'la chinoiserie', a noun which, following French rules of language, was used in a feminine rather than a masculine form.

Even more significantly, and less metaphorically, China's female role also emerged as it was effectively raped by Western powers. The Opium Wars, and the subsequent Japanese invasions, most effectively symbolized by the Rape of Nanking, were physical assaults on an unwilling China and helped to cement long-standing, and harmful, representations of China in a feminized gender role. The Rape of Nanking represented the extreme of Chinese feminization as it occurred at the hands of the Japanese, a historically smaller Asiatic power which had also been subject to Western intrusion. When perceptions of China eventually did begin to shift in the twentieth century, the contrast with those that came before was immense as the idea that

China could engender alarm among [American] voters came as an abrupt departure from the past. A legacy of missionary work and World War II efforts to rescue Chinese victims from Japanese aggression cast the Chinese as impotent – sometimes contemptible, sometimes piteous, always weak and dependent.[63]

2.4 The China Threat Discourse

The Respected China Discourse concerns a China that, although powerful and influential, presents no threat to the West due to its great spatial and temporal distance from it. The Disrespected China Discourses are spatially closer yet temporally distant as they represent an undeveloped China that is too weak to present a threat to the West to which it is subjugated. Speaking of American attitudes towards China in the 1950s, Nancy Bernkopf Tucker explains that Americans "might [have seen] Chinese rulers as venal and ruthless, but incompetence meant that China posed no challenge to the United States".[64] The China Threat Discourse, however, presents us with a China that is spatially and temporally close, a China which is, far from being emasculated, strong and assertive, and a China which is not only capable of governing itself, but is so successful at doing so that it provides an alternative model of governance to that of the West, as the Beijing Consensus gains ever more currency over the Washington Consensus with many newly developing nations.[65] This is occurring with Chinese governmental emphasis "that every country should find its own development path".[66] It will be demonstrated in the examination of CNOOC's failed bid for Unocal how this puts the CCP at odds with the Bush administration which, through the *National Security Strategy*, stressed values intrinsic to *universal* good governance. Thus, the CTD emerges from the West's newfound inability to effectively control China as it has traditionally been able to do. Consultant Therese Geulen states, "China has a lot of self-esteem, believing its place in the world should be at the top", and this causes concern in other countries.[67]

The CTD is also dependent upon a more precise Self than preceding China discourses as the US, rather than a broad notion of 'the West', becomes even more central to it. The chronology of the China discourses helps to illustrate this development. Although this book emphases thematic analysis over temporal analysis, there have been several periods where thematic elements have been closely tied to specific themes. The CTD is a contemporary discourse which is strongly informed by Sino-American relations. A timeline examination will illustrate that although different discourses rise and fall there is rarely a clearly defined event which marks a shift between one discourse and another. Instead, discourses tend to overlap one another. This is true of the CTD as although it has been most acute and has gained the most traction in popular discourse over the past three decades, its origins can be traced back to the nineteenth century at a time when Disrespected China Discourses were most prominent.

The CTD represents an image of a China which is antagonistic to the United States and which is mobilized against US interests. Essentially, and in keeping with central realist tenets, the China threat is a worst case scenario based on perceived antagonism of China towards the US. Susan Shirk writes that

> The question of whether China is a threat to other countries cannot be answered just by projecting China's abilities – its growth rates, technological advances, or military spending – into the future as many forecasters do. Strength is only one part of the equation. Intentions – how China chooses to use its power – make the difference between peace and war.[68]

Because it is impossible to 'get inside' another actor, analysis of intentions is risky if not impossible, and this book does not engage with it for this reason. However, having undertaken significant discourse analysis of issues surrounding China's rise, this book claims that there is a discernible American voice which argues that China poses a threat to it. The CTD diverges from other China discourses as it combines perceptions of increasing Chinese capabilities with those that China harbours malicious intentions towards the US. Previous China discourses included some which showed admiration for China and its capabilities and others which showed disdain and pity for China. Prior to the CTD, however, perceptions of China as a strong actor had not been associated with China as a threat to Western interests. When China was perceived as a strong and capable actor, the dominant discourse was the Respected China Discourse which portrayed a strong, but wise and benevolent China. When China was perceived as malicious, the dominant discourse was the Disrespected China Discourse which portrayed a cunning and ruthless, yet weak and ineffectual China. The CTD merges aspects of these two discourses to portray a China which is cunning and ruthless as well as strong and assertive. The CTD is predicated on many constitutive elements and notions of the Yellow Peril, the Red Menace, and China's economic and military growth will be explored in turn.

2.4.1 The Yellow Peril

Examination of Chinese discourses has thus far overlooked the racial aspects which inform such discussions. The racial aspect of Sino-Western relations is most clearly articulated in Yellow Peril rhetoric: "a set of paranoid race-fantasies" which concerns the construction of Western racist attitudes towards Asians.[69] The Yellow Peril is important to the CTD as it Others China by describing a racial hierarchy in which the 'yellow' Chinese are subordinate to Western 'whites'. 'Yellow' is used to position and devalue the Chinese as a group because "Light skin ... is almost universally valued".[70] To this thinking, the further from 'white' an individual or group is considered to be, the less civilized they are, which has historically resulted in their more disdainful treatment by 'civilized' whites. Such treatment was particularly pronounced during the era of segregation in the US when "A prominent Japanese observer wondered 'If Americans can regard Negroes as inferior, how do they really regard Asians?'".[71]

It should be acknowledged, however, that this is not purely a Western trend as there are historical precedents of associating undesirable traits with darker skin tones in many non-Western societies as "Hierarchies based on light skin are prevalent in Hindu cultures of India (Hall, 1995) and in other Asian and Hispanic cultures as well".[72] In discussing racial issues both inside and outside China, Frank Dikotter states that

> The polarity between white and black, derived from a differentiation of social classes and a particular aesthetic value system, was projected upon the outside world when the Chinese came into contact with alien groups. Black symbolized the most remote part of the geography of the known world.[73]

Desire for lighter skin tone in China is a noticeable phenomenon and "Asians spend an estimated US$18 billion a year to appear pale" because "lighter skin is a way of identifying with societies considered to be highly developed".[74] It is interesting to note however, and in keeping with their initial reverence of China, that in early contacts with China, Europeans ascribed no colour to the Chinese. The process of yellowing occurred tangentially with the increasing perception of the Chinese as backward as and less developed than the West, but as the twentieth century progressed, Chinese were increasingly seen as non-white or 'yellow'.

Although the Chinese would employ 'yellow' rhetoric, such as the cult of the Yellow Emperor, to buoy the legitimacy of the CCP, this was done as part of an effort to reappropriate the term 'yellow' to challenge the racist connotations which had become inherent in it.[75] However, despite such efforts, 'yellow' continues to be a derogatory term and has been utilized as such through many historical interactions between China and the West. This is central to the Yellow Peril as the inexorable links to racist ideas towards Asia help to enhance the threat to the West. Not only is this a peril, but one in which 'yellow' traits (e.g. overpopulation, cunning, non-Christian values) help to augment it. How, then, did the Yellow Peril originate, and how does it impact on the CTD in relation to US interests?

More than a racial trope, the Yellow Peril is

> a comprehensive discursive system with specific characteristics: the belief in the moral and spiritual degeneracy of Asian people; the fear of blending a superior race with an inferior race; the effect of Asian economic competition, and the threat of military invasion from Asia.[76]

The Yellow Peril originated in nineteenth century America with the first mass immigration of Chinese to the US.[77] Although the West showed a historical fascination with China and Westerners had been successfully able to integrate themselves into Chinese society, the Chinese showed little reciprocal interest in the West. Whereas there was a significant number of Europeans dedicated to Chinese studies and travel, Chinese did not begin migration of any real significance until the nineteenth century.[78] Although Chinese emigrated to many

Western countries, the impact of Chinese migration on perceptions of China was particularly important in the United States where Chinese 'coolies' and railway workers helped to define the image of China to stateside Americans. 'Pidgin English' and images of launderers and rickshaws became associated with the Chinese in America who were not only seen as different, but, in keeping with the infantilized theme, inferior as well. Chinese were perceived to be

> physically small, dirty, and diseased. In manner, they were allegedly humble and passive, but also sneaky and treacherous. They supposedly all looked alike and were depraved morally, given to theft, violence, gambling, opium, and prostitution.[79]

Mark Twain was one of the earliest American journalists to write about the landed Chinese and despite his attempts to provide balanced insights his reports reflect the bigotry of the time. More egregious examples of Chinese racial stereotypes in American literature could be referred to but a brief focus on Mark Twain is useful as

> Although Mark Twain's depictions of the Chinese are not always free from contemporary racial stereotypes, they are more sympathetic than what was typically portrayed in the popular media ... when Twain observed the Chinese, he was in fact examining the American character.[80]

His reports exaggerate stereotypes which were associated with the Chinese regarding differences in language, appearance, and mannerisms, and Twain's writings both served to demean the Chinese as well as highlight the unjust persecution which the Chinese suffered from bigoted America. For instance, he writes that Ah Sing "had in his store a thousand articles of merchandise, curious to behold, impossible to imagine the uses of, and beyond our ability to describe" but he also offered Twain "small, neat sausages" which Twain did not eat as he feared "that each link contained the corpse of a mouse".[81] Twain also remarked on Mr Hong Wo's (a Chinese lottery runner) "faultless English" as Wo explained the lottery system to Twain, where "Sometime Chinaman buy ticket one dollar hap, ketch um two tree hundred, sometime no ketch um anything; lottery like one man fight um seventy – maybe he whip, maybe he get whip himself, welly good".[82] Much in the way that the infantilized China discourse presents an image of the Chinese as fundamentally inferior to Westerners, such portraits by esteemed writers like Twain present the image of Chinese in America to be clownish yet interesting, but less than human.[D]

Upon landing in the country, the Chinese did not assimilate into American society. This was due, in large part, to purposeful exclusion from social life by white Americans. There was little Chinese integration or cohabitation into white America and Chinatowns began to surface. Nativist Americans (whites

---

D   See also: Ambrose Bierce, Bret Harte, Margaret Hosmer, Atwell Whitney, and Robert Woltor.

of European descent) began to resent the Asian influences on American culture and society and these Chinatowns acted as both a bastion for the Chinese themselves and as an area of quarantine for the reviled immigrants from whom whites could be safe.[83] However, although the Chinese influence on cities such as San Francisco and Los Angeles, as well as much of frontier America, was unwelcome by many white Americans, the influx of Chinese immigrants was not initially seen as perilous to the US even though these immigrants were seen as undesirable and were often subject to harsh persecution. Jonathan Spence notes that

> As the Chinese fanned out from San Francisco into new kinds of work in the mines and on the railroads, they moved from being objects of amused curiosity into targets of sarcasm, economic discrimination, legal harassment, and outright violence, sometimes ending in murder by lynch mobs.[84]

American paranoia led to the Chinese Immigration Act of 1882, which was the first policy enacted in America to exclude immigrants solely based on their race.[85] Thus, while an individual Chinese might have been disagreeable to white American sensibilities, there was great "nativist resentment toward a Chinese immigrant group" as Americans feared the yellow impact on their imagined white society.[86] This fear of a Chinese 'hoard' is echoed in modern-day threat assessments which use the size of the PLA to support the CTD.

Fu Manchu and the Yellow Peril

However prominent the perception of Chinese inferiority may have been in the late nineteenth and early twentieth-century Western discourses, it must be noted that it only represents one side of the Chinese exotic to Western minds. There was another equally different, but much more dangerous perception of China, best embodied by Sax Rohmer's Fu Manchu who, according to *Vanity Fair*, was "the most exotic and diabolic of contemporary villains in the annals of crime".[87] In total contrast to the Chinese as comically naive and unsophisticated, Fu Manchu was both evil and a genius, and could call on centuries of past Asian learning to bolster his nefarious plans for domination of the West. Essentially, Fu Manchu re-appropriated and redeployed an inverted Respected China Discourse. In *The Insidious Dr. Fu Manchu* Rohmer describes his character,

> Imagine a person, tall, lean and feline, high-shouldered, with a brow like Shakespeare and a face like Satan, a close-shaven skull and long magnetic eyes of a true cat-green. Invest him with all the cruel cunning of an entire Eastern race accumulated in one giant intellect, with all the resources, if you will, of a wealthy government, which however, has already denied all knowledge of his existence. Imagine that awful being, and you have a mental picture of Dr Fu Manchu, the yellow peril incarnate in one man.[88]

Fu Manchu's character is one of contrasts and oscillates between the defining characteristics of East and West. His physical description bears little resemblance to popular conceptions of the Chinese physique. He is tall and slim, and his features are compared to those of Westerners (i.e. Shakespeare) or products of Western thought (i.e. Satan). The description of his benefactor as being a 'wealthy government' is also not in keeping with the perceptions or reality of China at the turn of the 1900s. As well, the sinister image of Fu Manchu draws upon a great intellect. Although the Respected China Discourse rests on certain assumptions of historical intellectual prowess, by the early twentieth century scientific, technological, and industrial advances had ensured that the West had supplanted China as the centre of the intellectual world, making Fu Manchu's intelligence equally incongruous with the Asian associations attributed to him. Sax Rohmer was, perhaps, able to give credibility to his character by Westernizing Fu Manchu's strengths. Western dominance was also inverted by the Yellow Peril and Fu Manchu as his "near-total appropriation of socio-political and technological systems points to the negative capabilities of industrialization and modernization".[89,E] This negates the temporal distance and advantage, represented by modernity and development, the West enjoyed over China.

The real-life case of Dr Wen Ho Lee represents the personification of Fu Manchu in recent American affairs. Dr Lee was a Chinese American[F] nuclear scientist who was charged with mishandling nuclear secrets in 1999. Although he was later cleared of 58 of 59 felony counts by federal prosecutors, he was cast as Fu Manchu by American media[90] with the *Los Angeles Times* quoting FBI Director Louis J. Freeh as stating that Lee's "deliberate, unjustifiable, criminal actions put this country, and the world, at great risk".[91] The blogosphere also reacted to the media attention. In one thread, under the title "Wen Ho Lee should be SHOT!!", one response explicitly highlighted the role of Yellow Peril rhetoric surrounding the case and stated "Me thinks some folks seen too many Fu Manchu movies. Gee, could this be a 'Chinese thing?'".[92] The problematic racial aspects of Lee's case are explained by David Shih when he states that

> Critics of Lee's treatment denounced the government's characterization of the case as being inspired by the age-old 'Yellow Peril' fictions: because of his race, Lee never could be truly 'American' and so must be biologically inclined to serve Asian interests.[93]

Thus, the Fu Manchu-esque threat of Dr Lee stemmed from perceptions of his innate Oriental intelligence and his supposed inseparable allegiance to China, combined with ideas about his access to American industrial and scientific means. Because people read Dr Lee as a modern day Fu Manchu, and because he was perceived to be a threat to America by US elites, notably US Attorney

---

E   Echoes of the sinister appropriation of technology by the Chinese are found in US fears of Chinese cyberwarfare and are examined below.

F   Dr Lee was actually born in Taiwan in 1939.

General Janet Reno, Yellow Peril rhetoric worked to bolster the CTD which positioned him, and China, as an enemy.[94]

2.4.2 The Red Menace: China and Communism

The ideological gulf which exists between the US and the PRC provides a source of disquiet in relations between them. Although in reality neither China nor the US are the ideological paradigms they are sometimes perceived to be, the popular perception of the United States is that of a global champion of democracy while China is seen to be the last great Communist stronghold. Events in Beijing in 1989 underscore this. It is notable that while Communist regimes were falling en masse at the turn of the 1990s, the CCP worked against the tide of world history by violently rejecting calls for democratization – calls which even the Soviet Union answered. Such calls were, and are "anathema to Beijing".[95] Indeed, the American democratic ideal has been a primary engine of its foreign policy over the last century. Matthew Hirshberg states that "The cold war dominated American perceptions of the People's Republic of China during the 1950s and 1960s, and opinions of China were correspondingly negative".[96] In 2005 Liu Jianfei, director of the foreign affairs division of the Central Party School of the Communist Party of China (CPC), wrote

> The U.S. has always opposed communist 'red revolutions' and hates the 'green revolutions' in Iran and other Islamic states. What it cares about is not 'revolution' but 'color.' It supported the 'rose,' 'orange', and 'tulip' revolutions because they served its democracy promotion strategy

and the US strategy remains "to spread democracy further and turn the whole globe 'blue.'".[97] Indeed,

> there is an overwhelming tendency in [China threat] literature not to refer to the country as just 'China', or even the 'People's Republic of China', but repeatedly as *Communist China*.... It is a rhetorical debate concerning perceptions of China and is a game between the 'Red Team' (red for communism), and the 'Blue Team', pro-China-anti-America versus anti-China-pro-America players.[98]

Even as priorities have changed, American anti-communist values remain and have served to fundamentally shape its relationship with the PRC.

The success of the PRC in the mid-twentieth century is in large part a result of Western interference into China's internal affairs in the century prior. By the early twentieth century, in the wake of two Opium Wars and other Western misadventures, Chinese dynastic rule, which had overseen a dynamic China over several millennia, became stagnant and feckless, and in 1912 the Qing dynasty collapsed as Chinese rule was fractured.[99] Fairbank states that "The 37 years from 1912 to 1949 are known as the period of the Chinese Republic in order to distinguish them from the periods of more stable central government which came before and after".[100] Stability, if not immediate prosperity, only came with the Communist victory in the mainland in 1949. Although the success of the Communists in 1949 brought a level of stability to the domestic politics of China that it had not enjoyed under Republican or late dynastic rule, it completely upset China's place in the international system. In response to the 'loss of China' to Communism, the US joined with Australia and New Zealand to create the Australia, New Zealand, United States Security Treaty (ANZUS) in 1951 which committed the US militarily to Asia to oppose what Robert Menzies, the Prime Minister of Australia from 1949–1966, referred to as " 'the menace of Chinese Communism' and its 'primitive Marx-Engels gospel of aggression and violence' ".[101] Importantly, ANZUS helped to strengthen US regional commitments in the Pacific.

The Korean War is an important milestone in discourses linking the China threat with the Red Menace as "The Korean War ... changed earlier assessments and made a China threat tangible".[102] Mao's decision to 'lean to one side' and form a close alliance with the Soviet Union resulted in the Soviets' promise to "come to China's aid in the event of an attack by 'Japan or any other state which should unite in any form with Japan in acts of aggression' (a clear reference to the US)".[103] This Sino-Soviet Treaty of Friendship and Alliance was instrumental in enhancing perceptions of China's threat to the US as China became embroiled in the Korean War. Because the Chinese military, apart from its overwhelming size (echoes of a Chinese hoard), was easily outmatched by the US and its allies, China's primary threat to America stemmed from the fact that it fell under the Soviets' substantial defensive umbrella. Soviet assistance ensured China had access to modern materiel including fighter jets, and most importantly, China was gifted a nuclear deterrent. Because the Soviets did not engage directly in the hostilities, the major belligerents were China and the US who fought at the sharp end of a major Cold War episode. Because the Soviets and Americans never engaged directly in conventional warfare, the Korean War served to elevate China as a potent Communist enemy. As well, despite the fact that the Sino-Soviet alliance only lasted until 1960, at which point animosity between the powers led the Soviets to renege on their pledge to aid China,[104] the PRC's initial fraternity with America's primary adversary served to deepen Red Menace discourse surrounding China, and render it perpetually suspect.

The association between the Korean War and the China threat was robust and endured long after the cessation of hostilities. *M*A*S*H*, an American cultural institution of the 1970s which was released as a book, film, and television series, is representative of the way China's participation in the Korean War lingered in the American consciousness. While the book, by Richard Hooker (a.k.a. Dr Richard Hornberger), and film, by Robert Altman, used the setting of the Korean War to critique the Vietnam War, it is Larry Gelbart's television series which was most illustrative of the way American perceptions of the Korean War helped to further perceptions of a China threat. *M*A*S*H* supports televisual representations of Chinese, often by proxy through their North Korean allies, as malevolent and violent, yet simultaneously and conversely infantile and vulnerable. The only sympathetic Asian characters were Korean private soldiers and civilians who were subject to the inhumanity of the officer class and Chinese soldiers who feared Americans, yet longed for American care and medical attention. *M*A*S*H* also engaged in the discursive practice of infantilizing, feminizing, and dehumanizing most of the East Asian characters which helped to support notions of the Red Menace. *M*A*S*H*'s contribution to the perpetuation of the Red Menace is even more disconcerting when one considers that it purposefully attempted to undermine normative inequalities in America and sought to criticize the conservative US establishment. The extreme degree to which China was associated with the Red Menace discourse is evidenced by the fact that it was so unexceptional that an avowedly liberal institution such as *M*A*S*H* failed to subvert or even question it.

China's 'Red Menace' and the spectre of McCarthyism

While America faced increasingly fewer communist adversaries throughout the 1990s, the CCP stubbornly and aggressively retained its hold on power in China which ensured that the Red Menace remained a powerful tenet of the CTD. China's reaction to calls for democratization are explained by Roderick MacFarquhar who states that "No one [in China] wants to be the next Gorbachev because that will mean the end of the Communist Party".[105] In light of Chinese elites' seeming determination to maintain the CCP, despite the decrease in anti-communist rhetoric which accompanied the end of the Cold War between the USSR and the US, notions of the Red Menace remained in American discourse but were transposed from the Soviet Union onto China.

This transposition can be seen in the case of Dr Lee which illustrates how the Yellow Peril was enhanced by the Red Menace to situate China in a discourse smacking of mid-century American paranoia. Not only was Dr Lee an enemy because he was ethnically Chinese, but he was doubly an enemy because of the communist nature of the Chinese government. This case illustrates that even though the PRC's commitment to communist tenets and ideology continued to be eroded as its success in the global marketplace continued unabated[106] the US continues to position China as an Other through reference to its communist credentials. Despite the fact that US anti-communist rhetoric was relatively sparse by 1999, the year in which Dr Lee was imprisoned, if one were to read of Dr Lee's case in a 1950s context when rampant McCarthyism was a defining feature of American political life, his persecution would not seem incongruous. James Gibson explains that

> During the 1950s, the United States was undoubtedly a society characterized by considerable consensus in target group selection. The Communist Party and its supporters were subjected to significant repression, and there seemed to be a great deal of support for such actions among large segments of the political leadership as well as the mass public.[107]

China's persuasion to communism in 1949 coincided with the opening salvoes of Senator Joseph McCarthy's attack on suspected American communists through the House Un-American Activities Committee (HUAC) which would continue through one of the frostiest decades of the Cold War, until both McCarthy's increasing political isolation and premature death at the age of 47 put an end to these particular witch hunts. This did not, however, put paid to equations of communism as un-American. Despite its dilution in decades which followed, Red Menace discourse continued to be a potent component of the China threat as evidenced by US Representatives' calls to stop the sale of Unocal to a 'communist' country, and this will be explored in depth in the case study analysis. The invocation of the threat Chinese communism posed to the US in order to prevent a commercial market transaction is indicative of the resonance which remained with the Red Menace. The presence of McCarthyism clearly remains present as Representatives argued that Unocal's sale to CNOOC would be "pro-Communistic or unpatriotic".[108]

The 'Red Menace' and rapprochement

In 1969 President Nixon outlined what he saw as a pressing concern regarding China when he wrote in *Foreign Affairs* that "we do not want 800,000,000 [people] living in angry isolation. We want contact".[109] Henry Kissinger viewed China with caution and was quoted as saying in 1969 that although

> he had no quarrel with the desirability of reducing tension, he persisted in wondering whether an isolated China, so long as it caused no major problems, is necessarily against [US] interests.... A China that was heavily engaged throughout the world could be very difficult and a dislocating factor. Why is bringing China into the world community inevitably in [US] interests?[110]

However, the Sino-Soviet split represented an opportunity for America to reconsider its relationship with China and by the early 1970s Kissinger had seen the utility of forging closer ties with China, as Eisenhower and Dulles had in the 1950s,[111] and wanted a meeting

> so long as it included our basic themes: that we wanted to make a fresh start; that we would not participate in a Soviet-American condominium; that we would proceed not on the basis of ideology but on an assessment of mutual interest.[112]

Although the ideological divide remained, the US and China hoped to bridge it through rapprochement. However, neither China nor the US knew how to effectively deal with and interact with the other, and although the meeting between President Nixon and Chairman Mao initiated a process whereby China began to be reintegrated into the Western-centric international community, certain obstacles remained.[113] There remained significant impediments to any meaningful US-Sino cooperation as there was ineffective US-Sino understanding. Jonathan Mirsky, writing in July 1972, draws attention to the problem which perceptions, and their construction plays when he wrote that

> Since this Peking summitry furnished the big news in 1972 for Americans, it may seem curious that the Chinese seem unmoved by the Mao–Chou–Nixon meeting. On the contrary – the present China craze in this country constitutes merely the latest example of American ignorance of life in the People's Republic. Since my return to this country, the questions put to me reflect not the realities of China, but impressions filtered through baffled television reporters.[114]

Despite rapprochement, America seemed to remain wary of China. Mirsky goes on to state that

> Americans juggle two conceptions of modern China, which occasionally interlock, like those Hong Kong ivory balls, but usually fly off in different directions. Sometimes China is 'Chinese,' exotic but loveable, like pandas. At others we fear the Red/Yellow Peril, slopping over its borders while obscurely convulsed within. Depending on our needs, we have emphasized first one image, then the other.... The principal reasons for our mystification are lack of direct experience, [and] the misinformation provided by most modern China specialists.[115]

Although "The 1980s represented 'golden years' in Sino-US relations as prevailing American imagery became increasingly complementary", the oscillation between positive and negative perceptions of China ensure the Red Menace was reinvigorated in the late-1980s.[116]

The Red Menace and Tiananmen

Although Deng's stewardship of post-Mao China appeared to set it on a course of comprehensive rehabilitation and international integration, events in Beijing in June of 1989 suggested that China's normalization was a deception. Rather than comprehensive transformation, the Chinese government's reaction to protests in Tiananmen suggested that market reform was sought without any accompanying political reform.[117] The PLA crackdown on Tiananmen Square protesters reinvigorated notions of the Red Menace in popular Western discourse as China seemed to stand alone against the groundswell of democratization overtaking previously communist regimes on an otherwise global level. Oliver Turner writes that

> Despite apparent signals from China that it was now following in the footsteps of the West yet another Chinese 'revolution' had failed to conform

to American expectations. The imagined geography of Uncivilised China existed to Americans as starkly now as it had done a century earlier as it remained a nation and a people which lacked the imagined standards of the civilised Western world. It had taken just a few weeks for prevailing imagery of China to shift dramatically from overtly positive to negative but beneath that shift lay enduring and powerful continuities and commonalities.[118]

What the CCP referred to as "political turbulence" in later years, the *Washington Post*, in keeping with Western journalistic descriptions, called "one of modern history's most brutal crackdowns"[119] while the cover of *Time* magazine simply overlaid the title, 'Massacre in Beijing', on a photo of Beijing protesters surrounding a bloody corpse. Adding to critical perceptions of the CCP, the article "Crackdown in Beijing; Troops Attack and Crush Beijing Protest; Thousands Fight Back, Scores are Killed" by Nicholas Kristof, the *New York Times'* China correspondent, quoted a Beijing resident as saying "In 1949, we welcomed the army into Beijing.... Now we're fighting to keep them out".[120] In another article Kristof again promoted Red Menace imagery of corrupt CCP elite by contrasting Chinese double-digit growth in the 1980s with people's dissatisfaction with their leadership. He writes, "Throughout the country, the love, fear and awe that the Communist Party once aroused have collapsed into something closer to disdain or even contempt".[121]

The United States returned to a stance in which it perceived China, represented by the CCP, as a totalitarian Red Menace, a charge which occasionally emerges to this day and which was employed by several US representatives during the course of the CNOOC/Unocal affair. Two days after the crackdown, President George Bush even used the commemoration of the invasion of Normandy to admonish China when he stated "the momentous, tragic events in China give us reason to redouble our efforts to continue the spread of freedom and democracy around the globe".[122] Clearly, the bridge of rapprochement could not span the ideological divide between the US and China without great difficulty. Tiananmen upturned perceptions of China as a normalizing power of the 1980s as it once again became un-American, un-democratic, and a distinctly 'red' communist pariah.

2.4.3 *Economic resurgence: a new Asian assault*

The 1980s and 1990s were a time when the Asian Tigers began to challenge American economic hegemony. With a more interconnected and globalized world, systems of production became internationalized and companies like Toyota, Hyundai, Sony, and Samsung benefited enormously.[123] Japan and the newly industrialized powers such as South Korea, Singapore, and Taiwan were the primary players in the region and many Americans argued that their success was actually to the detriment of the United States. 'Buy American', "popularized in the U.S. as part of an effort to encourage American consumers to favour domestic products over imports", became a defining phrase of the 1980s as Japan's influence on the global economy, and its impact on the American market became alarming.[124] Buick and Chevrolet employed this strategy in an attempt to buck the success of their Asian rivals.[125] The backlash against this new Yellow Peril reached its zenith in 1989 when the Japanese Mitsubishi company bought several prominent New York landmarks including Rockefeller Center and Radio City Music Hall for $846 million.[126] The American angst at Japanese consolidation of US assets was made evident by Robert Cole: "The deal, which comes almost exactly 50 years after Rockefeller Center opened on Nov. 1, 1939, is only the latest instance of the Japanese buying a vital piece of the American landscape, from Hollywood to Wall Street".[127]

While the Asian Tigers represented the economic Yellow Peril during the 1980s and 1990s China was not regarded as a vital or dynamic economic powerhouse, but rather as the "sweat shop to the world".[128] However, with 10 percent growth per annum, China quickly found itself amongst the larger economic players, and the Asian Financial Crisis enhanced its reputation as a significant and versatile economy as China was not only able to weather the crisis which had plunged Japan into recession, but also helped to stabilize the region "by not devaluing its currency and by offering aid packages and low-interest loans to several Southeast Asian states".[129] There is an argument that the "good health of China's economy had a lot to do with cutting the Asian crisis shorter than it might have been" as China offered an alternative to the draconian measures offered by the IMF.[130,131] China was largely able to avoid the financial contagion which wreaked havoc on the economies of its neighbours and by the turn of the century it was poised to challenge the largest global economies. Economic growth became a central strategy of the CCP, especially in the wake of Tiananmen, "As the party realized that the performance-based legitimacy was the only hope for prolonging its rule".[132,133] By the turn of the twenty-first century, it was China, not Japan, which was the economic engine behind the resurgence of the Yellow Peril. Although in a radically different form the Red Menace resurfaced as China's economic prowess, rather than its ideological tenacity, began to be seen by some in the US as an affront to their national security.

William Reinsch, head of the National Foreign Trade Council, captured American fears of Chinese takeovers when he said, "You've got people in the Pentagon, and I gather other agencies, who don't like [Chinese investment into America], because the Chinese are bad guys – they don't know why, but by God, they're going to think of a reason".[134] The economic might of China therefore greatly contributes to the CTD. Carsten Senz, a Westerner working for China's Huawei company, states, "The fear of China can't really be pinned down to exact details.... The fear is difficult to grasp", but he believes as China's economic power can be easily understood, it is also quick to be blamed.[135] Opposition to Chinese investment into the US and fears about Chinese economic strength will be further explored in section 4.5.2.

2.4.4 China's military growth

China's military growth, increasing "at a pace that has closely mirrored its economic growth", is also essential to the CTD.[136,137] Anxiety surrounding China's military growth is expressed by a RAND report which stated that "China's defense spending has more than doubled over the past six years, almost catching up with Great Britain and Japan ... by 2025 China will be spending more on defense than any of our allies".[138] These anxieties are representative of US sentiments towards China which have been stable since the early 2000s.[139] While increased defence spending amongst allies tends to be welcome,[G] it is telling that "When the Chinese government released its latest defense budget, there was once again considerable angst in the United States and its East Asian allies".[140] It is clear that rather than viewing it as an ally who is willing and able to shoulder increased responsibility, the US views China as a competitor and is concerned about the "potential for China to mount a serious strategic challenge to the United States in Asia".[141] Threat perceptions of China's military spending emanate not only from the volume and increase in spending but also on the nature of PLA spending, and both are intrinsically linked to the CTD.

The volume of China's military spending

Official Chinese figures are generally greeted with suspicion because Chinese military spending is not transparent. However, the scale of Chinese military spending is reflected by the fact that US elites find even the underrepresented official figures alarming. Because China's military spending reports are opaque, the best insight offered into PLA spending is provided by the US Department of Defense's (DOD) *Annual Report on the Military Power of the People's Republic of China.* Pursuant to the FY2000 National Defense Authorization Act, these reports are compiled so that the US is kept apprised of the most significant developments regarding the PLA. The first reports were produced at the turn of the new millennium when perceptions of China's rise became entrenched as a potential future threat to US power. The first report, compiled in 2002, noted that China spent a publically reported total $20 billion but that real spending was likely $65 billion, "making China the second largest defense spender in the world after the United States and the largest defense spender in Asia".[142] Subsequent reports highlighted two trends which continued throughout the decade: double-digit growth of China's military and consistent underreporting on the amount of military expenditure by the Chinese government. For instance, the 2012 report noted that Beijing announced an 11.2 percent increase in annual spending, increasing its budget to $106 billion.[143] However, because "China's published military budget does not include several major categories of expenditure ... DoD estimates China's total military-related spending for 2011 range[d] between $120 billion and $180

G   The US regularly appeals to its NATO allies to be more equitable in their support for NATO and increase their defence spending from the 1.6 percent of GDP they spend on average to something closer to the United States' 4 percent (Kashi, 2014).

billion".[144] However, official Chinese figures might indicate an unwavering trend of an increased spending over the past decade, but they do not give details about what the military is spending money on. While the notable increases in Chinese military expenditure have caused alarm in the US, what has caused more disquiet is the secretive way China addresses its military spending as it leads to perceptions that this secrecy belies malicious intent.

The nature of China's military spending

The PLA has embarked on two paths of modernization and these offer insight into the identity politics which surrounds China's militarization. The first path emphasizes the development and acquisition of conventional military hardware, and the second path emphasizes asymmetric capabilities.

Conventional weapons and prestige

In its development and acquisition of conventional hardware China exhibits a desire to enhance its international prestige and redress perceptions of itself as weak and backwards. Although these perceptions of China were most pronounced in the nineteenth century, they persisted into the 1990s when China's weak military position forced it to capitulate to US pressure over Taiwan, "one of the remaining symbols of China's long period of weakness and dependence".[145] Although its leaders understand that the PLA is unlikely to be able to conventionally challenge the US military, China is nonetheless investing significantly in conventional weapons systems such as surface-to-ship missiles, fighter jets, and an aircraft carrier.[146] Gill Bates suggests that at the end of the last millennium China was a "non-*status quo*, dissatisfied power, determined to make up for its lost prestige and pride of place owing to the so-called 'century of shame and humiliation'".[147] Such notions persisted into the new century with others claiming that China's identity as a regional power has warranted its investment into conventional weapons as "military modernization is expected to enhance China's international prestige".[148] The strategy to enhance its reputation as a growing military power is best illustrated by the commissioning of the *Liaoning*, China's first aircraft carrier. This ex-Soviet ship has been refitted and deployed by the Chinese "to enhance China's national prestige".[149] Prestige, therefore, is central to China's acquisition of state-of-the-art conventional hardware as this allows China to play against identity type (e.g. weak, backwards, feminine) and to show its muscle without necessarily flexing it. Even if this conventional hardware is not deployed in combat, China perceives its ability to show it off legitimizes its claims to be a major power so that conspicuous weapons systems, such as the *Liaoning*, justify China's place at the tables of global power including the UN Security Council.[150] China is "still the only member of the United Nations Security Council never to have conducted an operational patrol with a nuclear missile submarine" and it is also "the only member of the UN's "Big Five" never to have built and operated an aircraft carrier" and it wants to compensate for this.[151]

This is, however, not to suggest that the modernization of China's conventional forces is purely ornamental. In its 2013 annual report to Congress, the DOD emphasized Chinese anti-access/area-denial (A2/AD), or what China refers to as "counter-intervention", as a prime area for US concern.[152] These include "short- and medium-range conventional ballistic missiles, land attack and anti-ship cruise missiles, counter-space weapons, and military cyberspace capabilities that appear designed to enable [A2/AD missions]".[153] PLA modernization also includes increasing its nuclear capabilities, advancing its long-range fighter capabilities, and increasing its power projection with the commissioning of the *Liaoning*.[154] The latter is also suggestive of China's intent to develop a capable blue-water fleet, and this is directly linked to perceptions of ES. Unlike Great Britain, the US, or Imperial Germany or Japan, China does not articulate its naval ambitions in terms of grand strategy although it

> may be drawn in this direction as its access to the world's commodities, markets, and thus oceans grows in strategic importance. Beyond trade, the increasing value and expected abundance of resources in and under the seas have added another incentive to gain and use sea power.[155]

The impact of the ESD on PLAN's modernization is evidenced because "Rising dependence on imported oil ... is fundamentally reshaping China's energy security strategy".[156] This trend is explored more fully in Chapter 4. In developing its capabilities, the PLAN need not enter into direct conflict with the US in order to render a significant blow to its navy. With a blue-water fleet China could reduce its dependence on the US in protecting global sea lanes of communication (SLOCs) and can thus reduce US leverage over China. Despite China's attempts to limit its reliance on SLOCs, "the sheer volume of oil and liquefied natural gas that is imported to China from the Middle East and Africa will make strategic SLOCs increasingly important to Beijing" making conventional blue-water capabilities essential to China.[157]

Asymmetrical warfare: undercutting the US

While the development of conventional forces reinforces Chinese prestige, the impact on its capabilities, with regard to challenging the US, is more marginal. The significant challenge China's forces pose to the US is their increasing asymmetric capabilities; capabilities meant to "complicate" the US "operational calculus".[158] Because China understands that its forces cannot compete with those of the US on a level playing field, China is developing its forces in a way which tilts the playing field in its favour and its A2/AD strategy is illustrative of this. The American security establishment has demonstrated concern that

> the PLA has turned to unconventional 'asymmetric' first-strike weapons and capabilities to make up for its lack of conventional firepower, professionalism and experience. These weapons include more than 1,600 offensive

ballistic and cruise missiles, whose very nature is so strategically destabiliz-ing that the U.S. and Russia decided to outlaw them with the INF Treaty some 25 years ago.[159]

Indeed, General Chen Zhou, the primary author of recent Chinese white papers, explains "the best use of our strong points is to attack the enemy's weak points".[160] While China uses conventional weapons to challenge the stereotypes of backwardness and weakness associated with it, its promotion of asymmetrical warfare inadvertently promotes Yellow Peril notions such as China-as-cunning because "China's history is replete with examples of the successful use of asym-metrical war, where wisdom rather than valour was used to subdue the opposing forces".[161] The endorsement of asymmetry by the PLA harkens back to the wisdom of Sun Tzu and is redolent of Fu Manchu underhandedness where, according to Qiao Liang, senior Colonel of the PLA, "the first rule of unre-stricted warfare is that there are no rules, with nothing forbidden".[162]

Thus, even more than its growth, it is the manner in which China's military is being modernized which is unsettling for the US. Despite modernization of rel-atively conventional technologies (i.e. missiles, fighter jets), Chinese A2/AD represents the most pressing threat to the US as smaller Chinese forces could countermand US supremacy. Easton explains that

> China has also developed a broad array of space weapons designed to destroy satellites used to verify arms control treaties, provide military com-munications, and warn of enemy attacks. China has also built the world's largest army of cyber warriors, and the planet's second largest fleet of drones, to exploit areas where the U.S. and its allies are under-defended.[163]

China's growing cyberwarfare capabilities have gained particular notoriety in America because if properly utilized these forces could circumvent the need for physical conflict and fundamentally undermine US military efficacy and its raison d'être. China need never again kowtow to the presence of a US carrier group in the Taiwan Strait if the PLAN is able to render that group ineffective by disrupting its communications or targeting and guidance systems.

Moreover, it is argued that the PLA regularly uses this technology off of the battlefield with regard to information gathering and "the Obama administration ... accused the Chinese military of attacking US government and defense con-tractors' computer systems. Some recent estimates believe that over 90% of cyberespionage in the US comes from China".[164] US officials claim that China's cyberespionage activities extend to the public sector and that Chinese cyber spies now pose a "quiet menace to our economy with notably big results.... Trade secrets developed over thousands of working hours by our brightest minds are stolen in split seconds and transferred to our competitors".[165] Indeed, on 19 May 2014 the US took the unprecedented step of accusing members of the Chinese military of conducting cyberespionage "marking the first time that the United States has levelled such criminal charges against a foreign country".[166] This highlights US attempts at vilifying China as a shadowy and underhanded opponent content to ignore the conventions of appropriate competition.

This section has sought to demonstrate that it is the nature rather than the overall increase in China's military spending which is most illustrative of how the US perceives China as a growing military threat. Because the US reads China as a strategic competitor, overall PLA growth is perceived as a threat to US capabilities. However, a closer examination of China's military spending demonstrates how US perceptions of a China threat rest on two assumptions. The first emanates from the investment into conventional weapons by the PLA. It is argued that the main function of this hardware is to bolster China's international prestige and to play against the traditional Chinese identity type as weak and backwards. The second threat perception stems from Chinese ventures into asymmetric capabilities. These capabilities bolster traditional perceptions of Chinese identity type as underhanded and cunning and unwilling to meet the US on a level playing field. Chinese cyberwarfare is most illustrative of this as China has come under great scrutiny and has been blamed for unfair practices by the US at a time when the US faces serious criticisms of its own cyber activities. Thus, threat perceptions which are linked to China's military spending are wide-ranging and often contradictory and they are mobilized in different ways by the US. For instance, American concerns over Chinese ambitions towards Taiwan and other South China Sea territories often invoke notions of PLAN's increasing conventional capabilities so as to portray China as increasingly strong, assertive, and bullish. However, American concerns over China's role in US cyber insecurity are articulated in a contrary way so that China is portrayed as evasive and devious. By highlighting these contrasting images of China, the US accomplishes the singular goal of portraying China as a growing military threat, and this is illustrated in both official and non-official American discourse.

2.5 Conclusion: the China threat

The use of discourse analysis in this chapter is essential to understanding the perceptions which have led to the particular view of China as threatening by some in the US. Chinese economic growth throughout the last three decades has occurred in spite of its differences with the US. Although greater economic integration helped to build and maintain a Sino-American relationship, distrust and political jockeying between the two powers makes the relationship uneasy. The cooperation and growing interdependence between China and the United States during the 1980s was an unprecedented time in their history. However, any promise as to the outright cessation of hostility between them was dashed by the PRC's crackdown on student protests, and Deng's support of the crackdown, in 1989.

Tiananmen provided a glimpse into the fragility that characterized a country governed by a rather closed-off socialist party in the midst of economic reform and 'opening up'. However, despite the political ramifications that followed the Tiananmen crackdown, China's growth continued. Although Tiananmen suggested that "the Communist Party had been losing its grip on the country"[167] and that an internal crisis was occurring within China, it was also of grave concern to the US as it reaffirmed the "the ideological incompatibility of China with the Western value system".[168] Had China been perceived as a 'responsible stakeholder' in an international system defined by Western values, fears of its growth might have been eschewed, but perceptions invariably arose that China was "not being a good citizen of the international community and not contributing to global public goods".[169] As US perceptions increased that these values were not shared, China's growth became menacing. China's rise impacted uniquely on the sole superpower of the United States whose hegemonic status has become challenged by China. Because the quantitative indicators of China's rise, such as its economic growth, do not necessarily suggest that China's rise must be perceived as threatening to the West, the qualitative constructs of the China threat become especially illuminating.

China's growth is a concern precisely because it is *China* which is growing. The once unfeasible fears of the Yellow Peril and the Red Menace have transmuted into a discourse where China may realistically threaten the West because of its enhanced capabilities. A spate of clashes between an increasingly assertive and capable China and the United States in the 1990s and early 2000s serve to illustrate this point.[H] Realist-oriented concerns of balance of power and relative gains, which I argue remain central to much US strategy, are exacerbated by the ideational differences between the United States and China. In essence, issues which would cause security concerns when viewed through realist optics in normal circumstances (e.g. the military build-up of China) are intensified due to the CTD.

Although Western perceptions of China have varied wildly, they have most always been critical in recent history. China may have been respected by the first Europeans to visit it, but this respect did not last long. Early Westerners revered China as being ahead of its time, but also perceived it to be equally eccentric. Western philosophical thought and scientific understanding also went through a period of extraordinary maturation which, by Western standards, eclipsed Chinese intellectual endeavours. Through epistemic hegemony the West was able to subjugate China which became perceived as feminized, infantilized, and generally regarded as something not to be concerned about. However, as China has found its footing throughout the twentieth century, and has embraced its strategy of growth which has carried it into the twenty-first century, the elements of Chinese culture and society which contrasted and clashed with those of the West (but were of no concern to the West due to China's inability to propagate them) are now perceived to pose a genuine threat.

H See the 1996 Taiwan Strait Crisis, the Chinese reaction to America's bombing of its Belgrade embassy in 1999, and the Hainan Island Incident of 2001.

Notes

1 Hansen, Lene. 2006. *Security as Practice: Discourse Analysis and the Bosnian War.* New York: Routledge.
2 Ibid., p. 95.
3 "China's Threats". 2000. *Washington Post*, 23 February.
4 Wan, William. 2014. "As Budgets Soar, China Still Fears Its Military Isn't Growing Fast Enough". *Washington Post*, 7 March.
5 Samuelson, Robert J. 2008. "The Real China Threat". *Washington Post.* 20 August.
6 Barr, Michael. 2011. *Who's Afraid of China? The Challenge of Chinese Soft Power.* London: Zed Books Ltd., p. 3.
7 Shambaugh, David. 2013. *China Goes Global: The Partial Power.* Oxford: Oxford University Press, p. 13.
8 Shaw, Han-Yu. 2012. "The Inconvenient Truth Behind the Diaoyu/Senkaku Islands". *International New York Times*, 19 September.
9 Carpenter, Ted Galen, and James A. Dorn. 2000. "China: Constructive Partner or Emerging Threat?" *CATO Institute.* From www.cato.org/publications/commentary/china-constructive-partner-or-emerging-threat, 10 May 2000.
10 Shlapak, David A., David T. Orletsky, Toy I. Reid, Murray Scott Tanner, and Barry Wilson. 2009. *A Question of Balance: Political Context and Military Aspects of the China-Taiwan Dispute.* Santa Monica, CA: RAND Corporation.
11 Ji, You. 2007. "Dealing With the Malacca Dilemma: China's Effort to Protect its Energy Supply". *Strategic Analysis* 31(3): 467–489, p. 477.
12 Steele, A.T. 1966. *The American People and China.* New York: The McGraw-Hill Book Company.
13 Turner, Oliver. 2011. "Sino-US Relations Then and Now: Discourse, Images, Policy". *Political Perspectives* 5(3): 27–45.
14 Fairbank, John King and Merle Goldman. 1998. *China: A New History.* Cambridge: Massachusetts, The Belknap Press of Harvard University Press, p. 151.
15 Polo, Marco. 1997. *The Travels of Marco Polo.* Ware, Hertfordshire: Wordsworth Editions Limited, p. 186.
16 Fairbank and Goldman. 1998. *China: A New History*, p. 92.
17 Spence, Jonathan D. 1998. *The Chan's Great Continent: China in Western Minds.* New York: W.W. Norton & Company, p. 1.
18 Koschorke, Klaus, Frieder Ludwig, Mariano Delgado, and Roland Spielsgart. 2007. *A History of Christianity in Asia, Africa, and Latin America, 1450–1990: A Documentary Sourcebook.* Grand Rapids, MI: Eerdmans, p. 20.
19 Ibid., p. 21.
20 Trigault, Nicolas. 1942. *The China that Was: China as Discovered by the Jesuits at the Close of the Sixteenth Century.* Milwaukee: Bruce Publishing Co.
21 Koschorke *et al.* 2007. *A History of Christianity*, p. 20.
22 Fontana, Michela. 2011. *Matteo Ricci: A Jesuit in the Ming Court.* Plymouth: Rowman & Littlefield Publishers, Inc., p. 17.
23 Ibid., p. 36.
24 Fairbank and Goldman. 1998. *China: A New History.*
25 Fass, Josef. 1973. "Chinese Newspapers". In *Essays on the Sources for Chinese History*, edited by Donald D. Leslie, Colin Mackerras, and Wang Gungwu. Columbia, South Carolina: University of South Carolina Press, p. 222.
26 Koschorke *et al.* 2007. *A History of Christianity*, p. 20.
27 Mackerras, Colin. 2000. *Sinophiles and Sinophobes: Western Views of China.* Oxford: Oxford University Press, p. 17.
28 Ibid.
29 Du Halde, Jean Baptiste. 1776. *The General History of China: Containing A*

Geographical, Historical, Chronological, Political, and Physical Description of the Empire of China. London: John Watts, pp. 128–129.

30 Crozier, Justin. 2002. "A Unique Experiment". *China in Focus Magazine*, Society for Anglo-Chinese Understanding (SACU).

31 Kerr, David. 2012. "China and Inner Asia: New Frontiers and New Challenges". *The Forum: Discussing International Affairs and Economics* 2012(Summer): 21–28, p. 23.

32 Mackerras. 2000. *Sinophiles and Sinophobes*, p. xxii.

33 Conrad, Sebastian. 2012. "Enlightenment in Global History: A Historiographic Critique". *The American Historical Review*, 117(4): 999–1027, p. 999.

34 Fairbank and Goldman. 1998. *China: A New History*, pp. 66–67.

35 Ibid., p. 3.

36 Bekar, Clifford and Richard G. Lipsey. 2002. *Science, Institutions and the Industrial Revolution*. Burnaby: Simon Fraser University.

37 Fairbank and Goldman. 1998. *China: A New History*, p. 3.

38 Mackerras. 2000. *Sinophiles and Sinophobes*, p. 35.

39 Voltaire. 1766. *The Philosophy of History*. New York: The Citadel Press.

40 Voltaire. 2000. "Splendid Secular Governance". In *Sinophiles and Sinophobes: Western Views of China*, edited by Colin Mackerras. Oxford: Oxford University Press, p. 35.

41 Mendoza, Juan Gonzales de. 1853. *The History of the Great and Mighty Kingdom of China and the Situation Thereof*. Translated by Sir George T. Staunton. London: The Hakluyt Society, p. c.

42 Smith, Adam. 1776. *The Wealth of Nations*. Petersfield: Harriman House Ltd., p. 47.

43 Ibid., p. 47.

44 Ibid., p. 439.

45 Macartney, George. 2000. In *Sinophiles and Sinophobes: Western Views of China*, edited by Colin Mackerras. Oxford: Oxford University Press, p. 57.

46 Rowe, William T. 2009. *China's Last Empire: The Great Qing*. Cambridge, MA: The Belknap Press of Harvard University Press, p. 91.

47 Doolittle, Justus. 2000. "Chinese Parsimony". In *Sinophiles and Sinophobes: Western Views of China*, edited by Colin Mackerras. Oxford: Oxford University Press, p. 77.

48 Ibid., p. 79.

49 Beresford, Charles (Lord Charles William De la Poer Beresford (1st Baron)). 1899. *The Break-up of China, With an Account of its Present Commerce, Currency, Waterways, Armies, Railways, Politics and Future Prospects*. London: Harper & Brothers, p. 89.

50 Mackerras. 2000. *Sinophiles and Sinophobes*, pp. 112–113.

51 Ibid., pp. 111–112.

52 Walter, Richard and Benjamin Robins. 1974. *A Voyage Round the World in the Years MDCCXL, I, II, II, IV by George Anson*. London: Oxford University Press.

53 Walter, Richard and Benjamin Robins. 2000. "Jesuitical Fictions". In *Sinophiles and Sinophobes: Western Views of China*, edited by Colin Mackerras. Oxford: Oxford University Press, p. 47.

54 Ibid.

55 Fairbank and Goldman. 1998. *China: A New History*, p. 3.

56 Frank, Caroline. 2011. *Objectifying China: Chinese Commodities in Early America*. Chicago: The University of Chicago Press, p. 39.

57 Ibid., p. 28.

58 Ibid., p. 60.

59 Pendergrast, Mark. 1999. *Uncommon Grounds: The History of Coffee and How it Transformed Our World*. New York: Basic Books, p. 15.

60 Ibid.

61 Ibid.

62 Frank. 2011. *Objectifying China*, p. 166.
63 Bernkopf Tucker, Nancy. 2012. *The China Threat: Memories, Myths, and Realities in the 1950s*. New York: Columbia University Press, p. 44.
64 Ibid.
65 Ambrosio, Thomas. 2012. "The Rise of the 'China Model' and 'Beijing Consensus': Evidence of Authoritarian Diffusion?" *Contemporary Politics* 18(4): 381–399.
66 Dent, Christopher M. 2011. "China, Africa and Conceptualising Development Relations". In *China and Africa Development Relations*, edited by Christopher Dent. New York: Routledge.
67 Jansen, Klaus. 2013. "German Anxieties Over China's Rise". *Deutsche Welle*. 20 August. www.dw.de/german-anxieties-over-chinas-rise/a-16963665.
68 Shirk, Susan L. 2007. *China: Fragile Superpower*. Oxford: Oxford University Press, p. 10.
69 Seshagiri, Urmila. 2006. "Modernity's (Yellow) Perils: Dr. Fu-Manchu and English Race Paranoia". *Cultural Critique*, 62(Winter): 162–194, p. 163.
70 Harrison, Matthew S. and Kecia M. Thomas. 2009. "The Hidden Predjudice in Selection: A Research Investigation on Skin Color Bias". *Journal of Applied Social Psychology* 39(1): 134–168, p. 157.
71 Bernkopf Tucker. 2012. *The China Threat*, p. 58.
72 Harrison and Thomas. 2009. "The Hidden Predjudice in Selection", p. 157.
73 Dikotter, Frank. 1992. *The Discourse of Race in Modern China*. Stanford: Stanford University Press, p. 12.
74 Barr. 2011. *Who's Afraid of China?* pp. 105–106.
75 Dikotter. 1992. *The Discourse of Race in Modern China*.
76 Kendall, Timothy. 2005. *Ways of Seeing China: From the Yellow Peril to Shangrila*. Freemantle: Curtin University Books, pp. 28–29.
77 Wu, William F. 1982. *The Yellow Peril: Chinese Americans in American Fiction 1850–1940*. Hamden, CT: Archon Books.
78 Ibid., p. 12.
79 Ibid., p. 13.
80 Ou, Hsin-yun. 2010. "Mark Twain's Racial Ideologies and His Portrayal of the Chinese". *Concentric: Literary and Cultural Studies* 36(2): 33–59, p. 33.
81 Spence. 1998. *The Chan's Great Continent*, p. 125.
82 Ibid.
83 Wu. 1982. *The Yellow Peril*, p. 66.
84 Spence. 1998. *The Chan's Great Continent*, p. 123.
85 Kil, Sang Hea. 2012. "Fearing Yellow, Imagining White: Media Analysis of the Chinese Exclusion Act of 1882". *Social Identities: Journal for the Study of Race, Nation and Culture*, 18(6): 663–677, p. 663.
86 Ibid., p. 664.
87 Seshagiri. 2006. "Modernity's (Yellow) Perils", p. 163.
88 Clegg, Jenny. 1994. *Fu Manchu and the Yellow Peril: The Making of a Racist Myth*. Oakhill, Stoke-on-Trent, Staffordshire: Trentham Books Limited, p. 2.
89 Seshagiri. 2006. "Modernity's (Yellow) Perils", p. 162.
90 Shih, David. 2009. "The Color of Fu-Manchu: Orientalist Method in the Novels of Sax Rohmer". *The Journal of Popular Culture* 42(2): 304–317, p. 304.
91 Drogin, Bob and Eric Lichtblau. 2000. "Reno, Freeh Insist Wen Ho Lee Posed 'Great Risk' to U.S". *Los Angeles Times*, 27 September.
92 woodyi...@my-deja.com. 1999. "Wen Ho Lee should be SHOT!!". [Online] Google Groups. Retrieved 12 November 2013, from https://groups.google.com/forum/#!topic/alt.politics/tl6KRa8VK5o, 31 December 1999.
93 Shih. 2009. "The Color of Fu-Manchu", p. 304.
94 Suro, Roberto. 1999. "Reno's Upset With Belated Video Disclosure". *Washington Post*, 3 September.

95 Carpenter, Ted Galen and Justin Logan. 2008. "Relations with China, India, and Russia". In *Cato Handbook for Policymakers: 7th Edition*, edited by David Boaz, 549–559. Washington, DC: Cato Institute, p. 552.

96 Hirshberg, Matthew S. 1993. "Consistency and Change in American Perceptions of China". *Political Behavior* 15(3): 247–263, p. 247.

97 Nathan, Andrew J. and Andrew Scobell. 2012. "How China Sees America". *China–US Focus*. Retrieved 17 October 2013, from www.chinausfocus.com/foreign-policy/how-china-sees-america/.

98 Broomfield, Emma V. 2003. "Perceptions of Danger: The China Threat Theory". *Journal of Contemporary China*, 12(35): 265–284, p. 267.

99 Tarling, Nicholas. 1967. *China and Its Place in the World*. Auckland: Blackwood & Janet Paul Ltd., p. 7.

100 Fairbank, John King. 1983. *The Cambridge History of China*. Cambridge: Cambridge University Press, p. 1.

101 Kendall. 2005. *Ways of Seeing China*, p. 31.

102 Bernkopf Tucker. 2012. *The China Threat*, p. 44.

103 Bailey, Paul. 2001. *China in the Twentieth Century*. Oxford: Blackwell Publishers Ltd., p. 160.

104 Ibid., p. 6.

105 "Expert: China Resembles the USSR Right Before the Fall". 2012. [Online] *Business Insider*. Retrieved 18 March 2014, from www.businessinsider.com/china-expert-no-one-wants-to-be-the-gorbachev-of-china-2012–6, 24 June 2012.

106 Broomfield. 2003. "Perceptions of Danger: The China Threat Theory", p. 268.

107 Gibson, James L. 1988. "Political Intolerance and Political Repression During the McCarthy Red Scare". *The American Political Science Review* 82(2): 511–529, p. 512.

108 Schrecker, Ellen W. 1988. "Archival Sources for the Study of McCarthyism". *The Journal of American History* 75(1): 197–208, p. 197.

109 Warner, Geoffrey. 2007. "Nixon, Kissinger and the rapprochement with China, 1969–1972". *International Affairs*, 83(4): 763–781, p. 764.

110 Ibid.

111 Bernkopf Tucker. 2012. *The China Threat*, p. 56.

112 Warner. 2007. "Nixon, Kissinger and the rapprochement with China, 1969–1972", p. 765.

113 Mufson, Steven. 1999. "Zigzagging over China". *World Policy Journal*, 1999/2000, p. 98.

114 Mirsky, Jonathan. 1972. "China after Nixon". *Annals of the American Academy of Political and Social Science* 402 (China in the World Today (July, 1972)): 83–96, p. 84.

115 Ibid.

116 Turner. 2011. "Sino-US Relations Then and Now", p. 39.

117 Schuman, Michael. 2013. "The Chinese Communist Party's Biggest Obstacle Is the Chinese Communist Party". *Time*, 25 November.

118 Turner. 2011. "Sino-US Relations Then and Now", p. 40.

119 Wan, William. 2013. "Witnesses to Tiananmen Square Struggle With What to Tell Their Children". *Washington Post*, 2 June.

120 Kristof, Nicholas D. 1989. "Crackdown in Beijing; Troops Attack and Crush Beijing Protest; Thousands Fight Back, Scores are Killed". *New York Times*, 4 June.

121 Kristof, Nicholas D. 1989. "China Erupts … The Reasons Why". *New York Times*, 4 June.

122 Bush, George. 1989. "Statement on the 45th Anniversary of D-Day". In *The American Presidency Project*, edited by Gerhard Peters and John T. Woolley. From www.presidency.ucsb.edu/ws/?pid=17117, 6 June 1989.

123 Hayter, Susan. 2004. *The Social Dimension of Global Production Systems: A Review*

of the Issues. Working Paper No. 25. Geneva, International Labour Organization, p. 1.

124 Avraham, Eli and Anat First. 2006. "'I Buy American': The American Image as Reflected in Israeli Advertising". *Journal of Communication*, 53(2): 282–299, p. 295.

125 Ibid.

126 Cole, Robert J. 1989. "Japanese Buy New York Cachet With Deal for Rockefeller Center". *New York Times*, 31 October.

127 Ibid.

128 McMillan, Andrew Frew. 2011. "China's Role as 'World's Factory' Coming to an End". *Invest China – Special Report*. From www.cnbc.com/id/41035650, 6 February 2011.

129 Shambaugh. 2013. *China Goes Global*, p. 96.

130 Ikenson, Daniel J. 2006. "China: Mega-Threat or Quiet Dragon". *CATO Institute*. From www.cato.org/publications/speeches/china-megathreat-or-quiet-dragon, 6 March 2006.

131 Fernald, John G. and Oliver D. Babson. 1999. "Why Has China Survived the Asian Crisis So Well? What Risks Remain?" *International Finance Discussion Papers* Federal Reserve System 633: 1–34.

132 Xia, Ming. 2006. "'China Threat' or a 'Peaceful Rise of China'?" *New York Times*.

133 Kristof, Nicholas D. 1993. "China Sees "Market-Leninism" as Way to Future". *New York Times*, 6 September.

134 Blustein, Paul and Mike Musgrove. 2005. "U.S. May Scrutinize IBM's China Deal". *Washington Post*, 25 January.

135 Jansen. 2013. "German Anxieties Over China's Rise".

136 International Institute for Strategic Studies (IISS). 2013. "China's Defense Spending: New Questions". *International Institute for Strategic Studies*. From www.iiss. org/en/publications/strategic%20comments/sections/2013-a8b5/china–39-s-defence-spending–new-questions-e625, 2 August 2013.

137 Kazer, William. 2013. "China Forecasts 7.6% Economic Growth in 2013". *The Wall Street Journal*, 26 December.

138 Crane, Keith, Roger Cliff, Evan Medeiros, James Mulvenon, and William Overholt. 2005. *Modernizing China's Military: Opportunities and Constraints*. Santa Monica, CA: RAND Corporation.

139 Perlez, Jane. 2012. "Continuing Buildup, China Boosts Military Spending More Than 11 Percent". *International New York Times*, 4 March.

140 Carpenter, Ted Galen. 2013. "China's Military Spending: No Cause for Panic". *CATO Institute*. From www.cato.org/publications/commentary/chinas-military-spending-no-cause-panic, 4 April 2013.

141 Crane *et al.* 2005. *Modernizing China's Military*, p. xv.

142 Department of Defense (DOD). 2002. "Annual Report on the Military Power of the People's Republic of China". *Report to Congress Pursuant to the FY2000 National Defense Authorization Act*. Washington, DC, United States Department of Defense, p. 38.

143 DOD. 2012. "Annual Report to Congress: Military and Security Developments Involving the People's Republic of China 2012". *A Report to Congress Pursuant to the National Defense Authorization Act for Fiscal Year 2000*. Washington, DC, Office of the Secretary of Defense, p. 6.

144 Ibid.

145 Carpenter, Ted Galen. 1998. "Let Taiwan Defend Itself". *CATO Policy Analysis*. From www.cato.org/pubs/pas/pa-313.html, 24 August 1998.

146 O'Rourke, Ronald. 2014. "China Naval Modernization: Implications for U.S. Navy Capabilities – Background and Issues for Congress". *Congressional Research Service*. Washington, DC, pp. 1–202.

147 Bates, Gill. 1998. "Chinese Military Modernization and Arms Proliferation in the Asia-Pacific". In *In China's Shadow: Regional Perspectives on Chinese Foreign Policy and Military Development*, edited by Jonathan D. Pollack and Richard H. Yang. Washington, DC: RAND Corporation, p. 20.

148 Brown, Harold, Joseph W. Prueher, and Adam Segal 2003. "Chinese Military Power: Report of an Independent Task Force Sponsored by the Council on Foreign Relations Maurice R. Greenberg Ceter for Geoeconomic Studies". *Council on Foreign Relations*. New York, Council on Foreign Relations, p. 2.

149 Glaser, Bonnie S. and Brittany Billingsley. 2011. "Is China's Aircraft Carrier a Threat to U.S. Interests?" *Center for Strategic & International Studies (CSIS)*. From http://csis.org/publication/chinas-aircraft-carrier-threat-us-interests, 11 August 2011.

150 Ibid.

151 Easton, Ian. 2014. "China's Deceptively Weak (and Dangerous) Military". *The Diplomat*. From http://thediplomat.com/2014/01/chinas-deceptively-weak-and-dangerous-military/, 31 January 2014.

152 Cheng, Dean. 2013. "Countering China's A2/AD Challenge". *The Heritage Foundation*. From www.heritage.org/research/commentary/2013/9/countering-chinas-a2-ad-challenge, 20 September 2013.

153 DOD. 2013. "Annual Report to Congress: Military and Security Developments Involving the People's Republic of China 2013". *A Report to Congress Pursuant to the National Defense Authorization Act for Fiscal Year 2000*. Washington, DC, Office of the Secretary of Defense, p. i.

154 Ibid.

155 Gompert, David C. 2013. *Sea Power and American Interests in the Western Pacific*. Santa Monica, CA: RAND Corporation, p. 5.

156 Shambaugh. 2013. *China Goes Global*, p. 270.

157 DOD. 2013. "Annual Report to Congress: Military and Security Developments Involving the People's Republic of China 2013", p. 19.

158 Shanker, Thom. 2012. "Pentagon Tries to Counter Cheap, Potent Weapons". *New York Times*, 9 January.

159 Easton. 2014. "China's Deceptively Weak (and Dangerous) Military".

160 "China's Military Rise: The Dragon's New Teeth". 2012. *The Economist*, 7 April.

161 Sehgal, Amrish. 2003. "China and the Doctrine of Asymmetrical Warfare". *Bharat Rakshak Monitor* 6(1).

162 Ibid.

163 Easton. 2014. "China's Deceptively Weak (and Dangerous) Military".

164 Cordesman, Anthony H., Ashley Hess, and Nicholas S. Yarosh. 2013. *Chinese Military Modernization and Force Development: A Western Perspective*. A Report for the CSIS Burke Chair in Strategy. Washington, DC, Center for Strategic & International Studies (CSIS), p. 60.

165 Nakashima, Ellen. 2011. "In a World of Cybertheft, U.S. Names China, Russia as Main Culprits". *Washington Post*, 3 November.

166 Nakashima, Ellen and William Wan. 2014. "U.S. Announces First Charges Against Foreign Country in Connection With Cyberspying". *Washington Post*, 19 May.

167 Kristof, Nicholas D. 1989. "China Erupts ... The Reasons Why". *New York Times*, 4 June.

168 Xia. 2006. " 'China Threat' or a 'Peaceful Rise of China'?"

169 Etzioni, Amati. 2011. "Is China a Responsible Stakeholder?" *International Affairs*, 87(3): 539–553, p. 539.

3 Central discourses

The Energy Security Discourse

3.1 Introduction

Having examined the China Threat Discourse, this chapter will explore the Energy Security Discourse and examine what it is and how it has become a dominant reading of energy security in the US. While the chapter is primarily concerned with the ESD, as it is pervasive in broad US ES discourse, it should be noted that other ES discourses exist as well, even to the point where they rival the dominance of the ESD in other regions and countries.[1,2] This chapter will demonstrate that the ESD represents a conventional reading of ES based on notions of relative gains. The ESD, explored in detail in section 3.3.3, is a reading of ES which is preoccupied with short- and medium-term, predominantly negative security perspectives concerning the acquisition of non-renewable energy resources, with a focus on oil. This contrasts to competing ES considerations which may privilege longer-term positive security perspectives concerned, for instance, with renewable energy resources. In addition, material concerns are paramount for the ESD as issues of scarcity and cost, with respect to resources, are twinned with conceptions of security which prioritize anarchy, relative gains, and states as primary actors. Where orthodox IR approaches can only provide descriptions of ES approaches, poststructuralist discourse analysis will allow us to uncover the assumptions and frameworks which constitute different discourses to see how they are actually constructed and articulated.

Discussion of the ESD in this chapter will differ from that of the CTD in the previous chapter because the manner in which the ESD is articulated differs significantly to that of the CTD. While the CTD is discussed in terms of radical difference to pro-China or even China-neutral discourses, the competing ES discourses use similar language to discuss similar phenomena. The differences between ES discourses emerge through divergent value-sets which prioritize issues differently from one another within the same epistemological and ontological parameters. While China discourses reference diverse factors to support various conclusions and are often articulated in either/or dynamics (e.g. China as threat/China as opportunity),[3] conventional and alternative ES discourses reference the same factors which indicate there is consent on the constitutive elements, but not their priority of importance. For instance, it is accepted within all

ES discourses that hydrocarbon resources are non-renewable and that their exploitation is generally harmful to the environment. It can be seen how different values can vary in importance within different ES discourses as an alternative ES discourse concerned with environmental sustainability would prioritize the environmental aspect of hydrocarbons while conventional ES discourses will prioritize their scarcity, and each would take a different meaning of ES from their particular reading. So although both discourses agree on characteristics of the nature of non-renewables they diverge in the way they accord importance to those different characteristics.

Although the ESD will be explored by referencing texts like the CTD was, the way the CTD is expressed could appear to be more apparent than the way in which the ESD is due to the explicit articulation of discourse in the CTD. Whereas the textual analysis of the CTD is immediately apparent as the texts are forthright about their meaning, the textual analysis of the ESD will be more subtle as the texts *infer* certain assumptions and meanings, and it is actually these inferences, not the statements themselves, which are of greatest import to the analysis. The ESD will emerge through exploration into a more hidden set of epistemological assumptions making it more oblique. Essentially, the CTD study looks to examine how language is operationalized, while the ESD study will look at the overarching structures to which language refers.

Explorations of the CTD and the ESD demonstrate differences in how they acknowledge their epistemological parameters. When undertaking a critical deconstruction of the China threat, these parameters become very apparent as it becomes clear that there is a Self/Other dichotomy that is not based on any causal relationships to truth or foundational knowledge, but is rather discursively constructed. Poststructural examinations of China discourses uncover qualitative processes of linking and differentiation which place China at varying subjective distances from the Self. However, through reference to material 'facts' (e.g. resource depletion rates, geographical location of resources), ES discourses tend to disguise the positivist epistemological framework as their claims to causality and foundational knowledge are more resilient. The resiliency of the positivist epistemology stems from the fact that ES rests equally on the constitutive notions of security and energy, and this is significant because while security may be a notion prone to deconstruction as it is not independent of human agency, energy lends itself more easily to positivist knowledge claims as it is predicated on the existence of resources which are materially independent of human action.

Because ES is often communicated in empirical terms as quantitative elements become prioritized, the language used with regard to ES is often less argumentative than factual. For instance, while China is *argued* to be a threat by China threat theorists, the ESD phrases issues such as scarcity, costs, and depletion not as *arguments* but as *facts*.[4,5] Because language associated with issues of energy make common inference to 'facts' about the material realities of energy resources, reference to such 'truths' helps to camouflage the structure so that discussion of ES becomes one of why-questions focused on events within the structure rather than on the how-question of the structure itself.[6]

The poststructuralist approach is therefore vital because it can engage with the discursive assumptions upon which the ESD rests and which often remain hidden by the material 'facts' of ES.

It is, for instance, a fact that the burning of coal releases more harmful emissions than that of gas.[7] It is also a fact that if our current level of oil exploitation is not curbed, oil reserves will be depleted as oil "is a finite natural resource, subject to depletion".[8] It will be demonstrated that material realities, such as these, are appropriated through discourses to pursue different arguments and agendas relating to energy security. It might seem that these material constraints are non-negotiable and would expose faults with an ideas "all" the way down perspective when applied to ES.[9] This, however, conflates the issue. Although the material realities of energy may not be deconstructed or reduced to constitutive elements, the poststructuralist approach helps us to rephrase the question so that we can examine what these realities *mean*. This is especially pertinent when energy is read in a security context. The fact that a bullet can kill a person means nothing outside of its discourse and explorations into the physical nature of the bullet will reveal nothing. Equally, the depletion of hydrocarbons means nothing outside of its own discourse which renders explorations into the physical properties of particular resources no more useful than those into the nature of the bullet. However, exploring hydrocarbon depletion alongside a discursive construction which elevates this 'fact' as pertinent to issues of security can be very revealing.

Because the ESD is a rather narrow conception of broader themes, and as it is one discourse among others, in order to examine the ESD this chapter will first turn to the overarching notion(s) of ES so as to provide a base upon which the analysis can build. Section 3.2 will examine the context of ES by examining what 'energy' is and what 'security' is. This section will focus on description over analysis and will foreground the analysis of the security aspect of ES by giving an account of the material factors of energy upon which discursive judgements about their meaning are made. Post-positivist conceptions of security will also be explored as they will be utilized in later sections to highlight the discursive nature of different readings of ES. Section 3.3 will be more complex as it will examine how ES is valued and measured. The section argues that ES is based on four factors; availability, accessibility, affordability, and acceptability. These factors will be examined and then addressed in connection with official and non-official ES discourse in order to highlight the universal and diverging elements of ES in the broadest context to give the reader an overarching idea of what ES can be. Section 3.4 will build upon this by examining how ES is read. While section 3.3 provides a broad outline of what ES can refer to, without the means to investigate specific discourses which exist within the wider umbrella of ES, the ESD would remain ambiguous. Therefore, section 3.4 examines how discourses, including the ESD, can be read. Because it makes reference to the elements to which language refers, rather than reference to language itself, section 3.4 helps to illustrate how different the exploration of the ESD is than that of the CTD. However, the utility of a poststructuralist approach will be demonstrated to be as valuable in this analysis as it is in that of the CTD.

3.2 Approaching energy security: a contestable concept

Amongst those who study it, a common refrain regarding ES is that there is little or no agreement as to what it actually refers to.[10,11] Without critical insight, the notion can be deceptive as it may initially seem uncomplicated and theoretically undemanding as the term 'energy security' can appear to be self-evident. Felix Ciuta, based at UCL, suggests that

> Those who survey the literature on energy security would be tempted to conclude that there simply is no need to debate what energy security is, because we know both that energy is a security issue and what security is. Yet the relationship between the two is profoundly problematic and warrants close conceptual and theoretical scrutiny.[12]

Ciuta suggests that understanding what 'energy' is and understanding what 'security' is leads one into a false sense of confidence when addressing the composite issue of 'energy security'. Although one would be wise to adhere to his call for more attentiveness to discursive constructions, even Cituta's statement makes the assumption that notions of energy and security are self-evident and non-debatable which is itself overly simplistic. As they are rarely fully articulated, without discursive examinations, 'energy' and 'security' are rendered susceptible to superficial and underdeveloped readings which promote particular assumptions, and this compounds the difficulties in understanding ES as a term. Ciuta's claim that "we" know what security and energy are, is equally problematic as no reference as to the composition of the we-group is made, and this impacts on contextual understandings of the term. Ciuta's assumptions are, in fact, indicative of many assumptions of ES, and this indicates the large degree to which the ESD and its component parts, with its inherently conventional nature, has become entrenched as *the* approach to ES from Western-, particularly American-elite perspectives. The discourse analysis which follows will illuminate the idea that far from being unbiased and apolitical, inherent agendas and interests constitute and support the ESD. Although ES is a contentious term, and although values will differ between ES discourses, the primary goal of ES in all of its guises is the SOS of energy resources.

Because it is dependent on subjective notions of values and beliefs and goes beyond mere reference to material facts, understanding ES requires thorough discursive investigation. However, such investigation requires a framework within which to proceed, and the first step in providing this framework will be to articulate conceptions of what we refer to by energy and security. This will provide the context within which discursive examinations can take place in section 3.4.

3.2.1 Energy context: resource type

This section will begin the discussion of the context of ES by briefly discussing what energy is. Christopher Dent and Elspeth Thomson explain that "Energy is often referred to as a 'total field', being inextricably linked to economic development and welfare, security, trade, communications, business and many other aspects of human endeavour and activity".[13] Ugo Bardi highlights the importance of energy to our particular societal context by stating

> Energy is the key factor that drives the economy. Without abundant energy coming from sources other than human and animal muscles, the society as we know it would be unthinkable. Energy is needed to power all kinds of machinery, but also for the vital task of supplying the industrial system with the mineral commodities that make it function.[14]

While reference to the importance of the economy and industrial continuance is highly contextual, this statement does effectively position energy as a prime consideration in our post-industrial society. Yergin provides further context for the importance of energy when he writes that

> "There is no substitute for energy," Schumacher said in 1964, echoing Jevons, the nineteenth-century economist and celebrator of coal. "The whole edifice of modern life is built upon it. Although energy can be bought and sold like any other commodity, it is not 'just another commodity,' but the precondition of all commodities, a basic factor equally with air, water and earth."[15]

By splitting the concept of energy along conventional/non-renewable and alternative/renewable lines, and by outlining different forms of energy within each of these groups, the aim is to provide an understanding of the options available to actors concerned with acquiring their ES. What this ES refers to, however, requires a more thoughtful investigation than a mere description of available resources, and this is what sections in the latter part of the chapter are tasked with. The reference to resources below will provide the reader with an adequate overview of the elements which, through acquisition, help to constitute the ES strategies of countries.[A] Examination of figures will be well served by poststructuralist scepticism of knowledge claims as the type of data which is provided can be used to illustrate the possible weaknesses in positivist epistemic assertions. To provide an example, it might seem notable that global hydropower output grew by 4.3 percent in 2012, but this actually reveals very little about the reasons behind this growth. For instance, this fact alone makes no reference to China's important role in this figure. The story that China's hydropower output is increasing (and adding to the global total) because of its desire to utilize more renewable energy, and of the renewable resources available to it China might prefer hydro to wind power due to grid constraints[B] is not told by the above

A Alternative ES approaches can consider non-state actors, but due to the US-Sino focus of this book, states will be examined as the primary actors.
B The distances the generated power needs to travel to the markets is markedly less between hydro plants and coastal cities than it is for wind farms making wind-generated electricity more of a burden on the grid.

percentage. This story can only be uncovered through investigation, and the figure, although it may independently indicate change, can only be understood alongside the discursive investigation. This provides but one example, but it serves to highlight that the 'facts' are dependent on discursive factors for meaning. Nonetheless, these figures do illustrate energy usage patterns and growth trends which can suggest broader patterns of energy usage.

An understanding of energy is important because "energy itself has the potential to affect substantially the way actors and theorists think about security in general".[16] The adoption of conventional versus alternative ES accounts is dependent upon one's set of values and beliefs and this correlates to the priority one gives particular energy resources over others. Although the physical properties are important, the material understandings of energy do little to inform matters of ES in relation to contemporary geopolitics. To provide a context within which exploration of the discursive constructions of ES can take place this section examines energy types in order to be able to demonstrate later how the prioritization of certain resources both stems from and fosters particular readings of ES. Although this study focuses on oil security, the importance of oil as a resource cannot be fully understood without reference to the renewable and non-renewable alternatives which exist. Despite the fact that they are perceived as a response to conventional/non-renewable energies, alternative/renewable energies will be addressed first.

Renewable resources

Renewable energy refers to a resource "that is derived from natural processes (e.g. sunlight and wind) that are replenished at a higher rate than they are consumed" or resources which "are continuously replenished by the natural world".[17,18] Most basically, renewable energy is considered to be cleaner and more abundant than non-renewable sources, but less powerful and more expensive. Renewable energy, according to organizations such as the IEA and the DOE, includes "moving water (hydro, tidal, and wave power), thermal gradients in ocean water, biomass, geothermal energy, solar energy, and wind energy".[19,20]

Because renewable energies, by definition, do not rely on limited resources, but rather in harnessing energy that is, for practical purposes, limitless, contemporary discussions of energy, of which resource availability is an essential part, increasingly surround renewable energy sources.[21] The increase in importance placed upon renewable energy mirrors the decrease in the availability of non-renewable resources such as coal, oil, and gas.[22] While renewables are often seen as an answer to the most problematic issues associated with non-renewable resource exploitation (e.g. scarcity, pollution), it must be acknowledged they also face particular restrictions and despite the promise that they hold, there remain great impediments to the comprehensive implementation of renewable resources into our energy mix.[C] Due to the high capital costs of building renewable power

C   See: Christopher Dent. 2012. "Renewable Energy and East Asia's New Developmentalism: Towards a Low Carbon Future?" *The Pacific Review*, 25(5), 561–587.

plants as well as integrating their output into existing grids, heightened expense is also often associated with renewable energy.[23] This is due, in part, to the fact that our energy infrastructures have been established around non-renewable energy resources, which means that our infrastructure is not tailored toward renewable sources and this ensures renewables are not cost competitive with non-renewable energies. However, this cost competitiveness should not be read as a weakness of renewable energy, but rather a simple corollary of the dominant reading of non-renewable energy as 'orthodox' and 'normal'.

Because of the expense associated with it, the research and development (R&D) and implementation of renewable energies tends to be a reserve of the global rich, although industrializing countries are making inroads as although

> the OECD remains the main source of renewable power generation (77.5% of world total in 2010) … non-OECD growth has accelerated sharply since 2007 and has exceeded OECD growth in percentage terms in each of the past three years.[24]

Non-OECD growth doubled from 17.5 percent in 2007 to 36.3 percent in 2010 although they still only accounted for 0.6 percent of non-OECD energy consumption in the same year, while renewables accounted for 2.2 percent of energy consumption of the OECD countries.[25] Dent also highlights the growing importance of renewable energy in East Asia when he explains

> East Asia's fast depleting indigenous energy resources and the inherent risks of depending on foreign sources of energy make new alternative sources of energy that can be generated at home (such as wind, solar, hydroelectric) increasingly appealing. This partly explains why East Asia's renewable energy sector has expanded faster than any other region in the world over the last five to ten years.[26]

This is especially pronounced in China which had, by 2010, "developed the world's largest installed capacity for renewable electric power at 263 GW, the US coming a distant second at 134 GW and Japan third at 60 GW (REN21 2011)".[27]

The idea that renewable resources tend to be clean is a significant part of the debate which surrounds their implementation. The World Nuclear Association summarizes this point effectively when they state that "There is a fundamental attractiveness about harnessing [clean] forces in an age which is very conscious of the environmental effects of burning fossil fuels and sustainability is an ethical norm".[28] Because of issues of non-renewable scarcity and environmental degradation, the implementation of renewable energy has become a significant part of ES conversations. The IEA has set targets to increase renewable electricity generation threefold between 2009 and 2035 to make it a more viable alternative to conventional resources.[29] However, as renewable resources accounted for only 13 percent of total global primary energy demand in 2010, it is clear that for renewable energy, a divide remains between the conversion of promise into performance.[30]

Non-renewable resources

Non-renewable resources are "created over geological ages" so that once they have been depleted it would take millennia for stocks to replenish.[31,32] In a sense, non-renewables are the inverse of renewables as they are 'dirtier' and less abundant than renewables, but tend, with current technology, to be more efficient and less expensive. In addition, unlike renewable resources, which are seen as an alternative energy option, non-renewable resources are the standard upon which our modern energy infrastructure has been built.[33] Our reliance upon these resources began in the Industrial Revolution as suggested by William Stanley Jevons, who stated,

> Coal, in truth, stands not beside but entirely above all other commodities. It is the material energy of the country – the universal aid – the factor in everything we do. With coal almost any feat is possible or easy; without it we are thrown back into the laborious poverty of early times.[34]

Though our energy mix has diversified to include other non-renewable resources in the years since, the essence of Jevons' claim that our modern society is dependent upon fossil fuels remains true.

In 1973, the year of the OPEC embargo, total world energy demand amounted to 3740 Mtoe (million tonnes of oil equivalent), 94.1 percent of which was non-renewable and only 5.9 percent of which was renewable.[35] Because our total energy demand has increased, although their contribution to the total mix has lessened, our total reliance on non-renewable sources has increased and they still constitute the majority of our energy mix. As of 2012, the total world energy demand was 5239 Mtoe, 81.3 percent of which is provided by non-renewable sources.[36] The fact that energy is read in terms of oil equivalency is also a telling sign as to the manner in which reliance on non-renewable resources has shaped our conceptions of energy in general. Although sources such as uranium contribute to the non-renewable resource mix, non-renewables primarily refer to hydrocarbons and these "Fossil fuels currently meet 80% of global energy demand".[37]

The primary constituents of non-renewable energy sources include oil (conventional and unconventional), gas (conventional and unconventional), and coal.[38] The reserve to production (R/P) ratio of oil suggests we have roughly 45 years' worth of reserves left if they were to be produced at the current rate as we have estimated reserves of five trillion barrels of conventional and unconventional oil.[39] Our gas and coal reserves are more abundant and geographically widespread.

3.2.2 Security context: positive and negative security

The question as to what security means must be addressed because "This question seems to divide the discipline not only because security could be called an 'essentially contested concept', but also – primarily so – because it is the starting point of many other related debates" regarding the "who, what, where and how" of security.[40] Buzan highlights the role of security by stating that "Security is about the ability of states and societies to maintain their independent identity and functional integrity" and it involves "a substantial range of concerns about the conditions of existence".[41] The way security is read fundamentally influences the manner in which energy is read and realist/rationalist explanations do not offer the analytic scope to fully explore issues of security. The poststructuralist approach adopted in this book allows for fuller understandings to be explored beyond positivist confines. The post-positivist stance which has been adopted opens up the possibility for a much more immersive exploration of security which allows us to examine what actors perceive it to be. If, as Buzan and Waever claim, "security is what actors make of it", security must be addressed with such understandings in mind.[42] This notion is supportive of Ciuta's claim that security is contextual.[43] It must, however, be maintained that this book is not concerned with what security actually is, but rather how it is read by the American Self, and it will be argued that the US values negative security over positive emancipation.

While this book argues that the actions of US elites have been informed by realist/positivist tenets, it also argues that their actions can be examined through the adoption of a post-positivist approach to security. Although it offers great insight into security approaches, post-positivism has significant divisions within it. Perhaps the clearest division with regard to post-positivist assumptions of security can be seen between those (e.g. Ole Waever) who support the notion of negative security, popularly linked with the Copenhagen School, and the others (e.g. Ken Booth, Richard Wyn Jones) who support the notion of positive security, popularly associated with the Welsh School.[44] This division will be explored below, as section 3.4 will demonstrate that the ESD, through which the US reads issues of ES, is informed by negative rather than positive security.

Negative security

Although both positive and negative security approaches are concerned with the limitation of threat, they differ in "the way in which security has been conceptualised and how scholars and practitioners themselves place a 'value' on security".[45] Arnold Wolfers provides a conceptualization of security when he states it is "nothing but the absence of the evil of insecurity, a negative value so to speak".[46] Security as a negative value is thus something to be avoided and is a concept "that should be invoked as little as possible".[47] With a focus on non-renewable resources, conventional approaches read ES as a negative value as these "definitions of energy security focus on the continuity of commodity supplies" which implies that uninterrupted supplies are the norm.[48]

Security is predicated on speech acts by those who adopt the notion of negative security and by linking the word 'security' itself to any issue (e.g. energy security), that issue will become a security concern and will be linked to a negative value. Reference to ES with emphasis on values such as scarcity underscores the negative security readings of ES. Huysmans quotes Isin when he writes that " 'to act, then, is neither arriving at a scene nor fleeing from it, but actually engaging in its creation' (Isin, 2008: 27)".[49] Thus, invocations of ES made through the ESD actually help to construct it in a negative sense as is evidenced by President Obama when he stated that, "At a time of such great challenge for America, no single issue is as fundamental to our future as energy. America's dependence on oil is one of the most serious threats that our nation has faced".[50] While negative security may not offer an adequately holistic understanding of all security issues, it does describe the way ES is read within the ESD, and is illustrated by its location in Figure 3.1.

Positive security

While security is treated in a negative sense by those who invoke the ESD to explain China's ES policies, there are others who read security as positive as it "is a good which provides the foundation to allow us to pursue our needs and interests and enjoy a full life".[51] Although positive security readings of the CTD and the ESD within Western discourse do exist, they are few and marginalized, and they are also represented in Figure 3.1.

The Welsh School approaches the issue of security from a fundamentally different way to those associated with the Copenhagen School. To proponents of positive security, security is considered to be a positive value and actors are emancipated from *in*security.[52] Security, to put it simply, is a condition to be strived for. To proponents of positive security, there is a marked difference between security and power and order in as much as security can be a universal condition while power and order can only exist at some other's expense. The conceptions of relative gains of negative-value ES are dismissed in favour of an ES account which strives for absolute gains. To positive security proponents, rather than drawing it into conflict with the United States, China's growing energy needs could "moderate its international behaviour" as China's "reliance on foreign oil could facilitate its deeper integration into the international system".[53] Moreover, positive security theorists could argue that a growing focus on renewable resources could underscore an emancipatory approach to ES as scarcity would no longer define energy acquisition in relative-gain terms. In keeping, positive security adherents share a space outside of dominant discourses as they appeal to alternative ES approaches with their lack of ties to conventional concerns such as states, fossil fuels, and scarcity.

3.2.3 Links between 'energy' and 'security'

Having broadly outlined the issues above, it appears that links are visible between different types of energy and different types of security. We can see that a clear link between negative security and a focus on non-renewable resources can be seen as the issue of scarcity can lead those who engage with the issues to prioritize negative security values of competition, power, and relative-gains. Conversely, a link is also evident between positive security and a focus on renewable resources as the emancipatory nature of resources which are not hindered by availability and scarcity are allowed to emerge. Both of these perceptions will be examined in section 3.4 which analyses the manner in which ES is read.

3.3 Measures of energy security

Having outlined the resources which contribute to energy production, as well as perceptions of security upon which ES can rest, in order to understand what ES can mean we must examine how it is constructed and measured. Rather than examining how it is read by different actors to glean particular meanings, this section will provide a rough outline of ES which rests on common constitutive elements which highlight a common set of goals. These elements are scarcity, price, efficiency/reliability, and environmental impact, but will be referred to as resource availability, accessibility, affordability, and acceptability.[54] All ES strategies balance these different measures in order to acquire SOS.

This section will illustrate how the components of ES, examined in section 3.2, may be arranged so that the poststructural analysis can take place in section 3.4. This section will begin by outlining the four factors of energy acquisition which have emerged through an immersive reading of ES literature and which define all ES approaches. Because this book aims to go beyond explorations of ES in the abstract, the following sub-section will examine official and non-official US discourse to see how these measures are prioritized in practice.

3.3.1 Four factors of energy acquisition

While ES is a contentious term, common themes emerge in any reference to ES and this section will discuss four factors which impact upon all ES readings. The Asia Pacific Energy Research Centre (APERC), explored in sub-sections below, refers to these factors with the alliterative phrase of the 'four A's'. These are availability (geological existence), affordability (the economics of energy acquisition), accessibility (geopolitical concerns), and acceptability (environmental and societal influences).[55,56] While ES discourses vary in their priorities (explored in section 3.4) they are uniform in their ultimate aim; ensuring the energy necessary for continued survival. What survival is taken to mean (e.g. economic success, social equality, ecological sustainability), however, creates the schisms between discourses. Thus, the 'four A's' are predicated on the opportunities or impediments to actors' abilities to acquire their energy needs, or more specifically, what they *perceive* their needs to be. Significantly, this latter point is often overlooked and this project aims to redress this oversight through the poststructural analysis in sections below. ES can also be sought anywhere from the individual to global levels, but the essential sameness of SOS goals for all actors underscores an epistemic homogeneity between approaches.

For the sake of clarity, these 'four A's' will be outlined individually in this section although they are intertextually dependent upon one another in practice and a poststructural examination of ES must recognize that an examination of one factor must necessarily intertextually acknowledge the other three. For instance, notions of availability cannot be entirely dependent on the physical availability of resources and take no account of the geopolitical, economic, or societal constraints which may be imposed by the other factors. As well, although all four factors are vital, it will be demonstrated in sections below how the primary schism between conventional and particular alternative ES discourses is predicated on different valuations of accessibility and acceptability. Essentially, issues of acceptability are of much less importance to conventional ES proponents than are issues of accessibility which tend to be privileged. On the other hand, alternative approaches which emphasize environmental and social concerns underscore the acceptability of different resources to the detriment of considerations of accessibility. However, as the general context of ES, rather than particular discourses, is of primary concern here, such debates will be deferred to sections which follow, and investigation into the four factors of acquisition will begin with the availability of resources.

Availability

The physical scarcity of resources is the most pressing concern for ES as all other issues are predicated on resource availability. Environmental, economic, and geopolitical concerns are all secondary to the actual material existence of resources. For instance, without energy resources, there would be no threat of 'high politics' resource wars.[57] Renewable resources are not always available due to physical realities which can affect their regional abundance or scarcity. Thus, although availability is often associated with scarcity of fossil fuels, this should not be solely considered at the expense of non-renewable analysis which underlies alternative ES accounts.

Renewable resources are only available in limitless supply when considered in a global context. However, ES considerations are most often taken on regional and national levels, and like non-renewable fossil fuels, renewable resources are not available everywhere.[58] According to BP's Energy Review, hydropower currently contributes 6.7 percent to the global energy mix.[59] However, this percentage should not suggest that hydropower is a universally viable energy source when we consider it at a regional level as the water resources necessary for power generation simply do not exist in desert regions, for example. Similarly, landlocked countries have no access to tidal power, countries at extreme latitudes cannot utilize solar power due to lack of sunlight in the winter months, and other countries are geographically distant enough from geologic activity to be considered geothermally barren. The result is that where renewable resources are available they can be considered to be infinite, but where they are unavailable they cannot be considered at all.

Availability is an even more apparent consideration for non-renewable resources. As finite resources, fossil fuels will be unavailable if they continue to be exploited at a rate where they contribute significantly to our energy mix because if no change is made in the current global R/P ratio, our global oil supplies will be depleted within a century.[60,61] The result is that scarcity of non-renewable resources is considered in the short-, medium-, and long-term, with a focus on the short-term, and this will be examined in section 4.4. The R/P ratios of resources are used to indicate the longevity of their availability, and the most well-known 'peak' equation was given by M. King Hubbert who suggested that resource "production must meet a maximum and then fall into inexorable decline".[62] His techniques include analyses of past resource discoveries, an estimation of future resource discoveries, as well as a projection of future production.[63] Although there is much debate surrounding 'peak' equations, there is agreement that non-renewable resource exploitation cannot extend indefinitely and that "production will one day end because it is a finite resource".[64]

Affordability

Affordability is significant to any conception of ES because a resource may be physically available but be prohibitively expensive so as to be financially unavailable to an actor. The issue of affordability is predicated on two inter-related considerations; investment and supply.

First, investment is necessary for both renewable and non-renewable sources. Although they have become intrinsic to the contemporary energy structure, non-renewable resources required significant initial investments in R&D and exploration and production (E&P).[65] Non-renewable technologies continue to develop and the utilization of new techniques is expensive and only undertaken when the output makes initial investment economically viable.[66,67] Deep-water drilling and fracking, for instance, only became feasible methods of oil acquisition when the price of oil made initial investments worthwhile.[68] Conversely, these resources will remain scarce if the investment does not increase with resource exploitation. Using oil as an example, the danger that there will not be enough supply to meet global demand at reasonable prices emerges "not because America and the world are running out of oil, but because they are running out of investment".[69] Sheikh Yamani, Saudi Arabia's former oil minister, illustrated the importance of continued investment to affordability when he stated that "The Stone Age came to an end, not because we had a lack of stones, and the oil age will come to an end not because we have a lack of oil".[70]

Affordability has had serious consequences for the implementation of renewable energy as investment into renewable energy projects is often seen to be "too expensive to be carried on a corporate balance sheet (Kann, 2009)".[71] Even hydropower, which is the renewable source which has been most fully integrated into our energy mix, suffers as it is perceived to "produce rather expensive electricity" due to high capital costs.[72] Such capital costs are also associated with other renewables meaning that until investment brings the costs down, or non-renewable resources become prohibitively expensive so that their alternatives become more appealing, solar, wind, and hydropower will remain marginal.

Second, affordability is inexorably linked with resource supply. As the scarcity of a resource increases so does its value, which, in turn, makes increased investments into its E&P economically viable.[73] Peter Bijur, the former chairman of Texaco, summarized the link between scarcity and the cost of oil when he stated that "It is true that we have probably been able to harvest most of the proverbial low-hanging fruit. But the higher fruit coming within reach [because of economic viability] is equally plentiful".[74] The affordability of resources can also fluctuate and this can be illustrated by the volatility of oil prices which has led to significant upheavals in the market.[75] Volatility has been a defining characteristic of the oil market in recent years with the price of oil spiking and reaching near record levels at the end of 2009.[76] While volatility does not tend to be associated with renewables, Narodoslawsky *et al.* state that "renewable resources are faced with considerable economic disadvantages. These disadvantages come from the fact that under current conditions renewable resources ... tend to be more expensive than fossil resources".[77] Thus, considerations of affordability are intrinsic to considerations of both renewable and non-renewable energy acquisition.

Accessibility

Geopolitical concerns underlie many considerations of ES. However, while the other three factors impact on both renewable and non-renewable resources, accessibility is unique in that it impacts almost solely on non-renewable sources. Where physical and geographical realities can make resources abundant in some places and scarce in others, political considerations can have the same impact.[78] The political considerations for accessibility to resources are of a dual nature. Limitations on access to resources can be both self-imposed as well as externally forced upon consumers.

US imports of crude oil from Iran help to highlight the notion that limits can be self-imposed. While US imports stood at just less than 600,000 bpd in 1978, US imports plummeted immediately following the Iranian Revolution, and have stayed low since.[79] Thus, despite its enormous export potential, Iranian crude has essentially become inaccessible to the US because of political barriers.

Conversely, the OPEC embargo provides an example of the way in which resources are made inaccessible to consumers by external powers. The 'oil weapon' was successfully deployed in October 1973, when OPEC placed an embargo on the US and the Netherlands for their support of Israel in the Yom Kippur War. Despite the fact that "the United States and the Netherlands were the state targets of the embargo, all oil-consuming countries suffered as a result of the oil supply disturbances".[80] Thus, regarding accessibility, Yergin states that "The problem is not one of running out [of oil] ... the real risk to supplies over the next decade or two is not geology but geopolitics".[81]

Acceptability

Acceptability is a contentious measure of ES which is central to the divide between renewable and non-renewable resources. Essentially, consumers have to accept a resource in order for it to be viably implemented. The primary issues which impact notions of acceptability are environmental and social.[82]

Gas, NGLs, oil, and especially coal face criticism from those who claim their environmental impact is unacceptable. Environmental concerns have gained increased attention over the past few decades because the traditional hydrocarbon-rich energy mix we have come to depend upon is becoming increasingly questioned. While their use, if checked, can theoretically be controlled, "irresponsible exploitation of [non-renewable] resources could have complex and possibly irreversible effects".[83] However, notions of acceptability can also hamper the implementation of renewable energies as detractors argue that they are more "inefficient" than their non-renewable counterparts and are thus harmful to the economy.[84] Other arguments are quite emotive and subjective. The perception that some, especially those associated with right-wing politics in the US, "are attacking everything 'green' " and view non-renewables with disdain and even anger is a potent contributing factor to the perception of renewable energies as unacceptable as an alternative to traditional fossil-fuels, and this will be explored below.[85] However, arguments surrounding nuclear power or large hydroelectric projects[D] underline the point that "debates on social acceptance are not totally new to the energy sector" and they affect renewable and non-renewable resource exploitation alike.[86]

3.3.2 Invocations of energy security

Although they are often not explicitly referenced, the concerns of resource availability, affordability, accessibility, and acceptability help to create a common understanding of ES. Although it has been illustrated that "Energy security means different things in different places"[87] the fact that reference is made to ES by many different actors and has "entered the lexicon of all those involved in the energy industry", and beyond,[88] suggests that there are common understandings shared by those who invoke it as to what ES can mean. This subsection will explore invocations of ES in official and non-official US discourse in order to uncover these understandings and see how these 'four A's' emerge in practice. As ES means different things to different people, overlap will occur in many, although not all, invocations of the term, but by illustrating that there is significant convergence on particular constitutive elements, broadly agreed-upon ES notions will emerge within which we can explore competing discourses. An examination of official and non-official policymaking accounts will illustrate that considerations which surround the central notion of SOS include economic, security, and environmental concerns, while other, more peripheral considerations

---

D   The furore surrounded the construction of the Three Gorges Dam is notable (McGivering, 2006; Tie, 2012; Wines, 2011).

(e.g. sustainability, desirability) orbit these central notions.[89] An examination of official and non-official US ES will take place in order to demonstrate how different conceptions of ES might be arranged, as well as audience accounts to see how marginalized issues are addressed.

Official policymaking accounts

With the energy crises of the 1970s energy security became central and Kerr explains that "a lasting consequence of this era was to elevate energy to a defining element in the fourth [core] power structure – security".[90] This was especially pronounced in the formulation of US national security, and "It is no coincidence that concerns about the ebbing of American power and the idea that we were suffering some kind of national funk, or malaise, were most pronounced at just the time the country suffered an energy crisis".[91] The US policies that were formed in response to these crises shared particular priorities which have carried through subsequent administrations to this day, notably issues of accessibility, affordability, and availability.

ES began to garner major attention in 1973. President Nixon highlighted ES in an address in April of that year when he stated "America's energy demands have grown so rapidly that they now outstrip our energy supplies. As a result, we face the possibility of temporary fuel shortages and some increases in fuel prices in America".[92] These shortages began to occur that October as OPEC placed an embargo on its oil exports to the US, and the continuance of these concerns throughout the decade is evidenced by President Ford's statement in 1977 that "Energy matters retain their troublesome hold among the problems threatening the Nation's long-run prosperity".[93] Oil, which had become a mainstay of the post-war US energy mix, was no longer as cheap a commodity as it had been for the US, and every American president to follow Nixon would place a high priority on oil's importance to ES. The early 1970s is also the time when Hubbert predicted American domestic oil production would peak and this ensured that issues of availability, access, and affordability topped American ES agendas.

The differences between ES strategies espoused by presidents and their administrations help to indicate varying ES policies. For instance, President Nixon responded to the OPEC embargo with 'Project Independence', a goal for the US "to meet America's energy needs from America's own energy resources", and this scheme was continued under President Ford.[94] However, President Carter later lamented that

> We've never had an energy policy for our country, because for too long we had extremely cheap oil – a dollar and a half or so a barrel – and we didn't have to worry about it, and we became addicted to – a highly dependent society on – foreign oil.[95]

Although price controls were actually instigated by the Nixon and Ford administrations, Carter's presidency coincided with the Federal Reserve's largest-ever interest rate increase in October of 1979 which plunged the economy into recession.[96] President Carter became associated with high fuel prices and gas lines which effectively "marked the beginning of the end" of his presidency.[97] His creation of the DOE and focus on energy conservation came under fire by the Reagan administration which attributed many of the problems associated with price controls to him. President Reagan stated

> Here in America, in this administration, our national energy policy dictates that one of the government's chief energy roles is to guard against sudden interruptions of energy supplies. In the past, we tried to manage a shortage by interfering with the market process. The results were gas lines, bottlenecks, and bureaucracy. A newly created Department of Energy passed more regulations, hired more bureaucrats, raised taxes, and spent much more money, and it didn't produce a single drop of oil.[98]

President Reagan's energy imports were hugely affected by the Iran–Iraq war, when energy prices spiked, and events in the Middle East continued to impact upon the subsequent Bush administration.[99] ES played a central role in President Bush's decision to declare war on Iraq with a coalition in which "the considerable burden of the effort [was] shared by those being defended and those who benefit from the free flow of oil".[100] Imports, especially from the Middle East, continued to impact upon US energy policy through the 1990s and President Clinton highlighted concerns of access when he stated that "the Nation's growing reliance on imports of crude oil and refined petroleum products threaten the Nation's security because they increase U.S. vulnerability to oil supply interruptions".[101] Though his administration made attempts to improve ES through greater domestic oil production and increased efficiency, President Clinton welcomed increases in OPEC production later in the decade as he claimed it would help "provide greater balance between oil supply and demand", thus finding common ground between access and affordability.[102] Despite attempts by US elites to reduce regional dependence on the Middle East as an oil exporter, the region remained as vital to US ES at the end of the century as it had been in the 1970s. Furthermore, oil retained its central place on the mantle of US energy resource dependency.

President George W. Bush suggested that the United States would take a more assertive role in securing its energy supply when he stated that "If we fail to act, our country will become more reliant on foreign crude oil, putting our national energy security into the hands of foreign nations, some of whom do not share our interests".[103] Perhaps the President intended his reference to action to be purposefully vague, but arguments abounded that US intervention in Iraq in 2003 was driven by a mercantilist desire to secure American oil interests by preventing Saddam Hussein from "extend[ing] his influence over the world's largest source of oil, the Persian Gulf".[104] Although the veracity of such arguments remain opaque, it is clear that oil imports retained a central place in US conceptions of energy SOS in the wake of 9/11. President Obama, who inherited the

War in Iraq, was more vocal about securing US energy requirements through diversification and increased efficiency,[E] and in doing so also stressed ES to be an important aspect of US national security. In order to ensure ES, his policies have attempted "a comprehensive and sustained effort, with emphasis on boosting domestic energy production, increasing efficiency, and transitioning to cleaner energy sources".[105] Although this brief inventory of official policy towards US ES suggests a great variance in approaches, there is actually more in common between these strategies than that which divides them.

The common goals of ES become clear when one considers that each US administration has focused on the SOS to ensure adequate energy resources for the health of the economy. However, while there is agreement to the constitutive elements of ES strategy, the policies of different administrations suggest that their order of importance varied. President Nixon attempted to address the US energy shortfall which resulted from the OPEC embargo by increasing production through measures such as drilling in the outer continental shelf.[106] President Carter, whose administration was arguably the most concerned with addressing issues of energy and environmental acceptability,[F] who attempted to tackle the energy crises by implementing measures which would increase energy efficiency and "reduce demand through conservation".[107] Both of these strategies were mobilized by the SOS of oil. Similarly, President Bush was confronted with SOS issues when Saddam Hussein invaded Kuwait and threatened its oil exports. The Washington Institute states that President Bush's strategy was based on oil and that US "worries over Saddam Hussein invading Kuwait were the fact of his aggression, his threatening the world's oil supply and being able to dictate the prices of the oil supply, by taking over the oil in Kuwait, and by threatening the oil in Saudi Arabia".[108] Used as examples, Presidents Nixon, Carter, and Bush demonstrate that while the strategies they used to combat energy shortages diverged, there is consensus between them as to the fundamental threat posed to US national security by energy shortages. Although ES strategies have differed between administrations, official US conceptions of ES all share some core concepts which include an emphasis on securing the supply of imported energy (with a focus on oil) in order to maintain the economic health of the country and ensure national security, and this highlights the importance of availability, accessibility, and affordability. However, with the exception of the Carter administration, substantive strategies to address issues of acceptability, commonly linked with environmental goals, are conspicuously absent from many official US ES strategies[G] which retain their primary focus on SOS.

E An echo of Democrats as 'green' (section 3.4).
F Another association of Democrats to 'green' agendas.
G This is not to suggest that environmental protection has been absent from Executive platforms as this is not the case. However, I argue that there is a critical difference between ES and energy policy, and I argue that much of the reference made by US presidents to the environment falls under the latter category.

Non-official policymaking accounts

Having briefly examined some official US approaches to ES following the energy crises of the 1970s, this section will examine how the 'four A's' emerge in non-official accounts of ES and determine how they relate to notions of import levels, economic health, oil importance, and environmental concerns. This section will examine texts from prominent think tanks in order to address ES in US discourses.

The Brookings Institution, a prominent non-governmental think tank based in Washington, DC, provides a good example of a non-official US ES definition. The Energy Security Initiative at Brookings states that there are three substantive aspects to energy security which include strategic considerations, economic perspectives, and environmental considerations.[109] The strategic perspective addresses geopolitical concerns at "the intersection of politics and energy" to examine the vulnerabilities to the SOS which may harm US national interests.[110] The economic perspective is closely associated with the strategic perspective as it examines the links between the economy and access to energy. Finally, the environmental perspective focuses on the link between the demand of energy, both renewable and non-renewable, and impacts on the environment including greenhouse gas emissions and climate change. It is of interest that the first two substantive aspects differ from the last in that the economic and strategic perspectives are clearly centred on the issue of SOS while the environmental aspect is not. By outlining ES in this manner, Brookings directly addresses accessibility, affordability, and acceptability.

Although established to undertake research into nuclear energy for the US defence industry, Sandia National Laboratories (SNL) has undertaken broader research into issues of domestic energy supply. SNL's account of ES is unique as its perspective lies in the middle ground between public and private sectors. Like Brookings, SNL states that ES rests on three pillars; environment, economics, and supply security.[111] The SNL approach differs to that of Brookings as rather than complementary, SNL understands the relationship between the three components (i.e. access, affordability, acceptability) to be competitive. In 2011 SNL asked 884 energy professionals to prioritize each component in order of importance (out of 100) which resulted in "the mean allocation toward the goal of Energy Supply Security was 36.9, the mean allocation toward the goal of the Environment was 30.7, and the mean allocation toward the goal of Economics was 32.3".[112] This highlights the important point that although, unlike Brookings, SOS does not apply to the economic issue in the SNL report, like Brookings, SOS was considered to be the priority for ES. The SNL study also suggests that a significant proportion of those polled view ES and environmental issues as separate to each other.

This separation is supported by Bob Tippee who stated, following a 2012 conference by The Near East South Asia Center for Strategic Studies (NESA Center), an affiliate of the US National Defense University, that "Most conversations about energy treat the environment and security as discrete matters to be dealt with accordingly".[113] This disunity between notions of security and environment can confuse matters as the latter is commonly considered to be a part of ES. The NESA Center's conference, like Brookings' ES definition, supported an approach to ES in which the constitutive elements are more intertextual and complementary than competitive. The elements which were highlighted at the conference included affordability, durability of supply, diversification of source, sufficiency relative to demand, equity among nations, and relationships between nations, all of which support the central role of SOS to ES considerations. The three elements which were raised at the conference which were not directly informed by SOS included environmental acceptability, relationship with water, and the morality of consumption, but while the "discussion about energy security that encompassed elements such as equity among nations, water, and the environment was interesting … the topic didn't receive much attention".[114] Thus, the NESA Center adds further support to the notion that the primary consideration for ES remains SOS and while other issues (e.g. environment) may have an impact on ES, they remain marginal. Ultimately, NESA appeals to issues of availability over acceptability.

The National Conference of State Legislatures (NCSL), a non-profit and bipartisan organization which works with the United States' 50 state legislatures and the DOE, states that "Energy security refers to a resilient energy system", and this resiliency also focuses on economic disruptions, public health and safety, and the potential environmental effects of energy security disruptions.[115] Although the NCSL definition makes no explicit reference to SOS, there are implicit and strong references to it which "show how central the energy system has become to American citizen's way of life" which also adds weight to the centrality of SOS to US perceptions of ES.[116]

Finally, as it coined the phrase, the Asia Pacific Energy Research Centre refers to all of the 'four A's' of ES.[117] While it offers a valuable framework, the definitions of APERC's factors suffer from its privileging of fossil fuels which can be evidenced by the association between resource availability and geological limits. Although the SOS referred to by all ES definitions above stress the importance of oil, ES conceptions must accommodate both non-renewable and renewable energies. APERC's focus on geological limits should therefore be replaced with one on geographical limits as non-geological considerations need also to be addressed as they affect the scarcity of renewable sources (e.g. areas with lack of hydro or solar capabilities). Thus, a separate list of four factors (scarcity, price, efficiency/reliability, and environmental impact) could be more inclusive as they are synchronous with APERC's approach but are less restricted by conventional/non-renewable concerns and extend to alternative/renewable considerations. Therefore, continued reference will be made to the 'four A's', but this reference will assume that the additional factors suggest above are included.

Public/audience accounts

While the official and non-official policymaking accounts of ES provide helpful insight into how ES is perceived by elites, and they therefore provide insight into the elements most important to the construction of ES policy, it must be remembered that the discursive nature of such policy requires audience acceptance. This sub-section will examine how non-elites, or the recipients of ES policy, view ES. This is important because it highlights different ES accounts to those above. Because ensuring the SOS is the primary goal for ES policymakers, it stands to reason that the public (policy recipients) need not worry about the same factors because the preservation of national SOS is beyond their responsibility. Thus, while issues of acceptability may be marginalized by policymakers, this section will demonstrate how they are articulated by the public. Additionally, while policy-oriented ES accounts almost universally articulate energy in terms of non-renewable resources, there exists much greater debate about the utility of renewable resources in audience accounts of ES. While availability and accessibility do not factor as major considerations, much debate surrounds issues of acceptability as well as affordability.

It is notable that renewable resources are read as being 'green', and a clean alternative to non-renewables.[118] In addition, technologies which harness renewable energies currently tend to generate much less power than do their conventional rivals. Renewable resources have been marginalized by the public through readings of green technologies as being liberal, to which derogatory readings as renewables as feminine, ineffectual, and expensive have been attached.

As happened to China during the Enlightenment, renewable resources have been disparagingly feminized and the process has served to challenge their acceptability within US ES discourse. This marginalization through feminization occurs in two tangential ways. First, there is a direct link between renewable resources' ability to halt environmental degradation and gender as "Dual themes recur throughout the existing though limited literature on gender and climate change – women as vulnerable or women as virtuous in relation to the environment".[119] Regarding virtue, renewable resources are 'green' and have a close association with regeneration, sustainability, and 'Mother Earth', as there is a normative "assertion of women's key role in protecting the planet and its natural resources".[120] Dalby underscores the idea that nature has been read through the male gaze and he writes, "Where Mother Nature is rendered as feminine in gendered tropes that emphasize phallocentric prerogatives to power and control, international relations has perpetuated relations of domination to the exclusion of women and the continued degradation of nature".[121] With regard to vulnerability, renewable energies have been called on to mitigate environmental degradation which has a disproportionate impact on women, in both developing and developed nations, as "gender vulnerability is compounded by a loss of control over natural resources".[122,123]

The second way the feminization of renewable energy has occurred is through perceptions of power. While conventional resources have been granted a sense of masculinity due to their inherent 'power', renewable resources are perceived as passive and weak, and the car provides a potent illustration of this contrast. Although not representative of America as a whole, Steve Tobak, writing for Fox News, illustrates this mentality in the extreme when he writes,

> The car I want to drive doesn't come in electric. There are two kinds of people in this world: car people and everyone else. I am a car person. I'm very particular about how I get from point A to point B. I have no idea why. Maybe it's an extension of my male ego or a character flaw. Whatever. All I know is, it's a free country, it's my money, and the car I want to drive doesn't come in electric.[124]

To Tobak, the roar of a V8 easily overpowers the quiet hum of an electric motor. Critics argue that the electric motor has been side-lined by "physics and math. Gasoline contains about 80 times as much energy, by weight, as the best lithium-ion battery".[125] The electric car provides a graphic example of the feminization of renewable, green energies.[126] For instance, the electric car was initially associated with women because "Ease of operation made electric cars the favorites of female drivers" and companies marketed electric cars to women by "making their products look like parlors on wheels".[127] It is even argued that due to the greater range and speed, men and women both preferred gas-powered cars, but "the *ideology* of the gender-specific choice of vehicle delayed the development of a really comfortable gasoline car".[128] Gender norms were central to the femininity associated with electric cars as "women who swore by their silent, *underpowered* electrics held assumptions about female delicacy".[129] Even though the correlation of the electric car with femininity was initially based on gender norms, once the feminization of the electric car reached cessation, the electric car, as a symbol of weak, renewable technology, also served to feminize its 'green' associates.

There is a correlation of perceptions between renewable energy, female gender norms, weakness, and political liberalism, with a focus on the Democratic Party in the US.[130] Although sweeping statements as to the equation of liberals as feminine is unsupportable, women do have a pronounced association with the Democratic Party as "Not only are women significantly more likely than men to identify as Democrats, and less likely to identify as independents, but – with only slight variation – this gap is evident across all ages".[131] According to an ABC News/Fusion poll, Democrats also register higher concern for women's issues, such as gender equality in the workplace and gender equality in politics.[132] Moreover, the link between gender, gender norms, and party is not only one of self-association, but also one of perception as evidenced by much of the right-wing American blogosphere. While one post refers to "limp-wristed Democrats" and the "prototypical girlie men of the Democratic Party", another states that "Weak sissy men are more likely to support welfare state, wealth redistribution and Democrat policy".[133,134] This link between gender norms and party "has become a regular feature of electoral politics, and Republican presidential candidates from Ronald Reagan to George W. Bush have had considerable success framing themselves as the stronger, more manly candidate".[135] Therefore, "As the Democrats become more committed to, and defined by, a green agenda" this derogatory feminization of the Democratic Party entails the feminization of Democratic policy objectives, including the integration of more renewable energy into the American infrastructure, policies which were addressed in the official discourse subsection above.[136] Because "Republicans and Democrats seem to be living on different planets when it comes to how to meet U.S. energy needs",[H] they disparage each other's policies, and "Republicans overwhelmingly push for more oil drilling [while] Democrats back conservation and new energy sources such as wind and solar power".[137]

Thus, perceptions of barriers to renewable implementation by the public are significantly "Cultural and behavioural" as well as aesthetic and they therefore form impediments based on the issue of acceptability which is largely absent in official and non-official policymaking ES accounts.[138] However, issues of acceptability remain relatively marginal in discursive formations of ES at the policymaking level.

Convergence of ES accounts

By examining some official and non-official US policymaking, as well as public accounts of ES it is clear that common themes emerge. Issues of accessibility and affordability are almost universally applied to ES considerations with regard to policymaking while availability and acceptability are not. In this regard, acceptability appears to have the least traction of the 'four A's' as many official and non-official ES accounts even fail to make mention of it. Brookings provides a useful description of ES by stating that,

> Energy security is a major factor influencing how countries conduct their foreign, economic and international security policies. Major supplier countries with vast energy resources exercise more power on the international stage than ever before. Energy is a primary consideration in how large importers – in need of adequate, reliable, and affordable supplies of energy – make alliances, offer foreign aid, and otherwise conduct their foreign policy.[139]

This summary represents official and non-official conceptions of ES and illustrates the ultimate goal of SOS. Moreover, Brookings' reference to the power exercised by "supplier countries with vast energy resources" implies the importance of oil and gas as strategic commodities. The non-renewable nature of these resources infers that negative security approaches are best equipped to address SOS. While availability, accessibility, and affordability all impact on SOS it is notable that issues of acceptability really do not. Acceptability largely remains a function of unofficial public debate and has little impact on the central ES notion of SOS.

H This can also be evidenced by differences in ES policies of the Nixon, Ford, and Carter administrations.

3.4 How energy security is read: discursive formations

Having examined the resources and types of security which contribute to ES policies in section 3.2, as well as the measures of ES in section 3.3, this section will examine how ES is read in order to show what actors *mean* when they refer to ES. This section illustrates frameworks within which ES discourses can be represented, compared, and understood. As discussed in the beginning of the chapter, differences in ES discourses will emerge less through explicit textual reference than through reference to implicit assumptions of epistemic structures. These assumptions suggest particular policies which can be illustrated in the framework which reflect the differing valuations of constitutive ES concepts. Differences in ES policies both reflect and produce differences in discourse and value certain components other others and

> This partly comes from the fact that energy security has a rather elusive nature and it is highly context dependent. Still, the fact remains that governments see security of supply as a major objective for their energy policy. The fact that energy security is strongly related to other policy issues that concern the energy system (such as affordable energy and climate change and environmental policy) implies that it is important to study the energy security consequences of different development pathways.[140]

Figure 3.1 illustrates a spectrum which can be used to arrange these different pathways to explain how ES is read and how it results from different perspectives. This section will begin by building upon the measures of ES outlined in 3.3 by positioning these four factors against each other to see how different prioritizations can lead to different readings of ES. Examining the discursive connections between issues of availability, accessibility, affordability, and acceptability will provide the foundation upon which a more immersive exploration into discourse can take place. This discursive exploration will examine the differences between conventional and alternative approaches to ES and is illustrated by Figure 3.2. This exploration is important because the ESD is a specific conventional reading of ES, and in order to be able to fully explore the discourse the wider conventional context must first be understood. This context is best illustrated through contrasting it with alternative approaches. The section will then conclude with a precise examination of the ESD itself.

3.4.1 Perspectives of energy security

Different perspectives of energy security can be portrayed in a spectrum which accommodates even the extremes of ES approaches and which is illustrated by Figure 3.1.[1]

---

1   Adapted from Kruyt, Bert, D.P. van Vuuren, H.J.M. de Vries, and H. Groenenberg. 2009. "Indicators for Energy Security". *Energy Policy*, 37: 2166–2181.

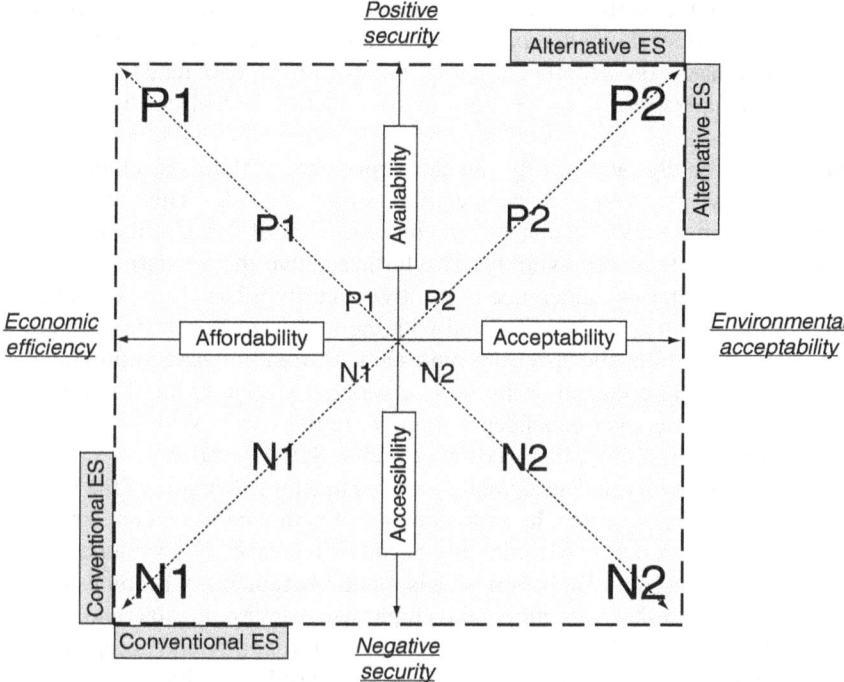

Figure 3.1 Framework of ES perspectives.

The perspectives suggested by this spectrum are compiled through reference to statements such as those outlined in section 3.3. This spectrum offers a roadmap to understand how ES is read rather than a concrete and prescriptive framework. The four measures of ES all inhabit the same plane in the spectrum and can be quantified by their locations on the different axes of the figure. The distance along a particular axis which a perspective is found indicates a greater degree of adherence to that particular ES measure.[J] Thus, these axes do not only represent simple binaries (e.g. convetional/alternative) but also complex and variable combinations of values. By locating where a particular cluster of measures lies within the spectrum, it is possible to draw some conclusions about particular ES perspectives.

Issues of accessibility become more pronounced the further below the *x*-axis they are, which represents increasing negative security perspectives the further from the *x*-axis a perspective is found. This contrasts with the measure of availability which becomes increasingly pronounced the further above the *x*-axis a

J   For instance, the larger P2s represent increasingly alternative approaches the further from the intersection they are located. The dotted line of the border also represents a permeable boundary to suggest that there is no defined limit to extreme ES approaches.

perspective is. This contrast occurs because the geopolitical concerns of accessibility become increasingly more immediate than the geographic concerns of availability the further down the axis a perspective is found, and vice versa. The increasing emphasis on negative security on the lower end tends to reflect increased regionalization as cooperation between actors is replaced with strategies of relative gains as perspectives emphasize perceptions of safety *against* insecurity. Conversely, availability, on the upper end of the axis, emphasizes positive security as these ES perspectives strive *for* ES. This engenders cooperation among actors so that geographic concerns of availability override geopolitical concerns of accessibility. The further above the x-axis an ES perspective is, the greater the adherence to positive security it has. Thus, unlike the regionalization which is often associated with the lower end spectrum of accessibility, the upper end of the spectrum represents "A trend towards multilateralism, market trust and cooperation in the international system [which] will most likely reduce concerns over dependence on other regions".[K,141] With its focus on availability and accessibility, the x-axis is central to SOS concerns.

Issues of affordability and acceptability are located on either side of the y-axis which, as will be demonstrated, helps to draw a distinction between conventional and alternative ES accounts. Affordability becomes increasingly pronounced the further left of the y-axis an ES perspective is found. Acceptability becomes more pronounced in the other direction and emphasizes environmental and other alternative ES concerns. There is conflict between affordability and acceptability as "the environmental benefits of renewable energy production have to be compared to their economic costs" which, at this time, is rather expensive, and this creates a clash between those who value economic performance over long-term ecological sustainability and those who value ecological sustainability over short-term economic imperatives.[142] Although the focus is on environmental concerns, the right hand measure of acceptability also accords increased importance to social considerations which are otherwise marginalized by conventional approaches.

The result is four 'storylines' which represent differing ES perspectives.[143] These storylines can be divided into two camps; two of which represent positive security views of ES and two which represent negative security views of ES. The first storyline, which emphasizes positive security in P1, also emphasizes economic efficiency and cooperation. Here, positive security is achieved through "convergence among regions, capacity building, and increased cultural and social interactions".[144] P1 is represented in the 'real-world' by the trend of increased M&A between major Western oil companies in the late-1990s and early 2000s[L] and the "consolidation wave that transformed the oil industry over the past decade".[145] This trend will be used in the next chapter to highlight the unique way the CTD worked with the ESD in influencing the way CNOOC's bid for Unocal was treated by US elites.

K   Negative security can, however, be associated with cooperation (e.g. strategic alliances).
L   See Chapter 4 for detailed discussion.

The second storyline, seen in P2, also emphasizes positive security. Like P1 this storyline emphasizes cooperation but it differs from P1 in that it emphasizes acceptability over affordability and alternative over conventional approaches to ES. Discourse in P2 is mostly represented by non-official and marginal groups such as the International Institute for Sustainable Development which suggests that "The relationship between the environment and globalization – although often overlooked – is critical to both domains".[146] The International Renewable Energy Agency (IRENA) "is an intergovernmental organisation that supports countries in their transition to a sustainable energy future, and serves as the principal platform for international cooperation" and effectively represents the P2 storyline in practice as it represents a group of countries striving *for* ES through cooperation in alternative means.[147]

The third storyline, N1, is aligned with negative security conceptions of ES and values affordability and accessibility. N1 is preoccupied with the regionalization of cheap and efficient non-renewable resources. While many official policies are representative of this storyline, President Nixon's Project Independence, referred to in section 3.3, is more illustrative than most as it sought to ensure that "Americans will not have to rely on any source of energy beyond our own. As far as energy is concerned, this means we will hold our fate and our future in our hands alone".[148]

N2 represents the final storyline which also represents a negative security reading of ES and highlights issues of accessibility, but also acceptability. President Carter's great variety of environmental and conservation programmes are representative of N2, as illustrated by his response to the oil crisis of the time where he stated, "As our nation increasingly turns to coal as a replacement for our dwindling supplies of oil and gas, we must be sure that we will not fall short of the goals we have established to protect human health and the general environment".[149] Although the interests which straddle the y-axis are traditionally incommensurable, President Carter attempted to bridge the competing issues of environmental concerns with economic health through focus on renewable energies like solar power, as well as a greater emphasis on conservation and efficiency. Thus, he sought to save the US *from* energy insecurity through alternative means. Although examples of N2 deployed in practice within the US are rare, it will be demonstrated in section 3.4.3 that examples of N2 are becoming less rare in Europe and especially East Asia.

Considering the whole figure, through examinations of official and non-official accounts it will be demonstrated that the ESD straddles the x-axis between N1 and P1, although its traditional perception of ES as a negative value ensures it is rooted in N1. More specifically, the next chapter will illustrate that the CNOOC/Unocal affair is highly representative of N1 storylines which result from the discursive links between the CTD and the ESD. Having examined a plane on which ES perspectives may be read, the next section will turn to the division between conventional and alternative ES considerations.

3.4.2 Conventional and alternative ES readings

Although ES discourses can be separated by a conventional/alternative divide, this divide does not represent a clear division between two cohesive and competing approaches to ES as there are many different ways conventional and alternative approaches could be employed depending on the contexts in which they are read, as well as by who is reading them. As they share a common epistemology, ES discourses, both conventional and alternative, can best be read through the same four components: resource type (section 3.2), security type (section 3.2), temporal perspective (section 3.4, below), and ultimately, ES accounts (section 3.4, below). The way these components interact is outlined in Figure 3.2 and will be expanded upon throughout the rest of this section.

While it is possible to arrange these four components in the abstract, the mobilized interests which generate and result from ES discourses can become clearer when they are critically explored in conjunction with a 'real world' case study with complexities of its own; indeed, there is a strong argument that energy security is entirely context dependent as "an exact definition of energy security … is hard to give as it has different meanings to different people at different moments in time".[150] The influence of dominant ES discourses could lead one to read particular contextual information in certain ways. For instance, the apparent clarity surrounding the enmity (perceived or otherwise) between the United States and China regarding energy is due to the fact that assumptions have been made as to what ES actually refers to and what it constitutes, and these assumptions stem from particular readings of particular issues. This book argues that dominant US ES readings, embodied by the ESD, are inclined

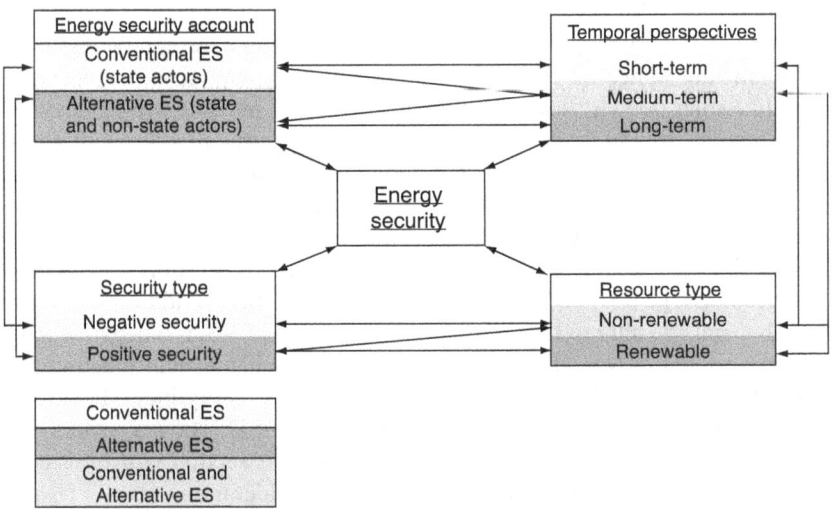

Figure 3.2 Conventional and alternative energy constructs.

towards conventional notions of ES and that that these readings are contextual and are magnified by the CTD.

Because the importance of context suggests that ES can mean different things to different readers, we must gather as holistic an understanding of all approaches as possible in order to study the ESD because despite its popularity, there are different readings of ES which challenge its assumptions. These alternative approaches champion different philosophies than do orthodox approaches to ES, and to understand the institutionalization of the ESD within the US establishment, these alternative readings, and how they result in often marginalized alternative approaches and discourses will also be examined. Because it is a term that has been bandied about and abused through generalizations by many academics and policymakers alike, those who invoke notions of ES often do not overtly articulate what they mean by it. It is therefore up to those who study it to question its meaning when ES is invoked as it is not a homogenous term.

Conventional readings of energy security

Daniel Yergin wrote that "The objective of energy security is to assure adequate, reliable supplies of energy at reasonable prices and in ways that do not jeopardize major national values and objectives".[151] Written at the turn of the post-Cold War era, his statement seems no less valid than it did two decades ago as "major national values and objectives", in conventional readings, are the eternal and primary concern of states, and states, in turn, remain the primary referents for the way ES is conventionally read. Indeed, all official accounts examined in section 3.3 placed an emphasis on the elements Yergin highlights.

Conventional readings of ES place a premium on a strong link between economics (affordability) and geopolitics (accessibility). Checchi *et al.* note that

> The literature is divided between those who interpret energy security from an economic perspective and those who stress its political and strategic side.... The literature is further divided between those who see the security of supply as exclusively related to energy and those who like to couple it with the environmental dimension.... Although there is no common interpretation, it is possible to identify a number of features that are always included, namely physical availability and prices.[152]

However, as conventional readings also prioritize the acquisition of non-renewable resources, the issue of scarcity (availability) also remains an overriding concern. Thus, although analysis of ES can extend to include issues of acceptability with regard to the environment, conventional ES concerns remain largely focused on issues of affordability, accessibility, and availability. These latter issues are addressed through two different acquisition strategies, the first being the neo-mercantilist/realist approach, and the second being the liberal political economic approach. In addition, as the focus of conventional ES is placed on the acquisition of non-renewable resources with an emphasis on oil, conventional temporal concerns are focused on short- and medium-term gains. These will all be looked at in turn.

Strategies for energy acquisition

The neo-mercantilist tradition of energy acquisition reads ES gains as zero-sum and can best be measured in terms of negative security which places it in quadrants N1 and N2 of Figure 3.1. The rationale of the neo-mercantilist strategy is that by owning the sources of energy, a state may insulate itself from shortages. Japanese involvement in the Second World War is often perceived to be driven by this rationale and "For some analysts, it is precisely such current fears of energy insecurity among Asian states which are likely to lead to conflict and war over enemy resources".[153,154] The neo-mercantilist tradition in the US is illustrated by the fact that energy independence has been invoked by six consecutive presidents because, according to Robert McNally, " 'Energy independence' is a popular mantra and guaranteed stump applause line".[155] Although the US employs neo-mercantilist strategies when convenient,[M] the CNOOC/Unocal affair demonstrates American unease at Chinese attempts to do so. Chapter 4 will explore how the US perceived Chinese ownership of Unocal assets, or "own[ing] oil at the wellhead", to be a direct threat to US interests.[156]

The liberal political economy tradition of energy acquisition represents a positive security approach to ES and exhibits faith in market forces to ensure SOS, and is represented in quadrants P1 and P2 of Figure 3.1. As American domestic oil production increases[N] the US has demonstrated a renewed confidence in free-market forces.[157] President Bush illustrated the importance of this strategy to US renewable technologies as it would enhance "international cooperation and technology investment".[158] However, while the US may extol the virtues of free-market mechanisms through political rhetoric on green technology, its actions suggest that it believes neo-mercantilist strategies retain an overriding importance as oil, and competitor importers, become positioned as security threats (e.g. CNOOC).

Temporal perspectives of conventional ES

Temporally, conventional ES is read in the short- and medium-terms. The short-term perspectives of ES emerge from focus on non-renewable resources with particular emphasis on fossil fuels. The conventional nature of short-term perspectives becomes evident as the goal of short-term energy security is "essentially to ensure that the national and regional economies are capable of managing a temporary disruption of supply" and national and regional economies are the

M   For example, President Reagan created the US Central Command to use the military to keep the Straits of Hormuz open in a crisis.
N   US oil production has increased from 5 mbd in 2008 to 7.5 mbd in 2013, 2.1 mbd short of its 1970 production peak.

preserve of conventional ES.[159] The safety of the global economy is not considered beyond the way in which it impacts atomized actors at the national level. Scarcity comes to the fore in this instance as energy users compete against one another for access or control of scarce resources such as oil. Thus, short-term readings of ES highlight issues of accessibility and affordability at the expense of availability and acceptability. ES readings with short-term perspectives are found in the N1 quadrant of Figure 3.1.

While short-term perspectives of ES are defined partly by national self-interest and relative gains, medium-term perspectives are less immediate and are more concerned with notions of sustainability. Because of the necessity to balance both immediate needs and projected requirements, the issues for medium-term security are demanding and complex. Importantly,

> there is a need to create the incentives and to foster the conditions for increased energy efficiency, so that the demand for energy is reduced, which is both necessary for climate change mitigation as well as for improving energy security.[160]

Reducing demand by increasing the efficiency with which resources are used is important to notions of medium-term ES as it changes the game from a zero-sum one into one in which gains become absolute rather than relative. Energy efficiency has been an increasingly important goal for energy policymakers and great gains have already been made including the Corporate Average Fuel Economy regulation in the US which is set to "save 1.8 billion barrels of oil over the life of cars and trucks sold between the 2012–2016 model years".[161] Especially with the fading popularity of SUVs, the image of the American 'gas-guzzler' is disappearing. The notion of efficiency has taken such root in the automotive industry that

> the best immediate hope for restraining the nation's fuel consumption might be some new vehicles that, although powered by conventional engines, run efficiently because they have been stripped of unnecessary weight, streamlined to move smoothly and equipped with gas-sipping engines.[162]

Medium-term perspectives expand on short-term perspectives and extend from the N1 quadrant into the P1 and N2 quadrants.

Alternative readings of energy security

While conventional ES tends to prioritize industrialized importer states with a focus on non-renewable resources, alternative ES approaches are wider and open to "development contexts" in that they examine the rich and poor, states and non-state actors, and examine renewable and non-renewable energies.[163] Human security, which has gained increasing attention in the post-Cold War and post-9/11 world, is representative of alternative approaches which claim conventional accounts are not illustrative of the 'real-world' as they "illustrate a limited problem-solving orientation that leaves the basic structure of world politics untouched".[164] Dalby argues that alternative ES approaches challenge conventional notions because they "have broadened the concept of security even further. In doing so, they have drawn on senses of the term 'security' beyond the specifications conventionally used in international relations".[165]

To alternative ES theorists, conventional accounts tend to act like Mercator projections and exaggerate Western problems at the expense of the rest. For instance, the Oil Crises of the 1970s, while disruptive, would not register as crises for undeveloped actors as "the citizens of the developed world enjoy a level of energy security of which the poorest of the people in the world can only dream".[166] When viewed through optics which prioritizes issues of human security, alternative ES is read as something entirely different to conventional approaches. While Yergin's definition may work with reference to industrialized nations, it has little relevance to someone whose energy security constitutes gathering enough wood fuel or other biomass to be able to provide heat for the day.

However, alternative ES can also be read in Western contexts, and alternative readings are often based on a prioritization of non-renewable resources as alternative approaches tend to stress the acceptability measure of ES readings and these readings populate the P2 quadrant in Figure 3.1. Whereas conventional readings of ES prioritize short- and medium-term concerns regarding the acquisition of non-renewable resources, alternative ES looks to longer-term considerations, often with a focus on renewable resources. These longer-term considerations are often inspired by concerns of availability as geological peaks are emphasized in these ES readings. Informed by alternative ES approaches, Adam Sieminski states, "Understanding the implications of long-term trends in global energy supply and demand is critical to any formulation of energy policy".[167] Although longer-term readings of ES do focus on geological availability, they also note that availability is discursively linked to accessibility and affordability. Dannreuther stresses the political and economic factors which enter into 'peak calculations' when he states that

> whenever the peak is reached, there is the need to be prepared, through the necessary research and investment, to ensure that this momentous transition away from a dependence on fossil fuels can be made without engendering severe economic and political conflicts.[168]

Despite the catastrophe which will occur if a solution is not found to the energy insecurity which results from ongoing dependence on non-renewable resources, states and IOCs remain incredibly reluctant to invest in the technologies and policies which would make such a transition possible due to the economic unattractiveness of such an exploration. Thus, longer-term energy security perspectives remain marginalized by medium- and short-term perspectives which are central to the ESD which will now be explored.

3.4.3 *The Energy Security Discourse*

Having explored how ES is read in a general context, as well as outlining more specific readings of ES by illustrating the divide between conventional and alternative readings of ES, this section will conclude by explicitly articulating what the ESD is. Although this chapter briefly outlined the ESD in the introduction, and has made inferences to it throughout, this concluding subsection will examine each of the four factors of ES to overtly outline how it is positioned. Taking the framework of the conventional/alternative divide highlighted by Figure 3.2, the ESD is effectively illustrated in Figure 3.3.

The ESD represents a thoroughly conventional account of ES and has become the dominant discourse of ES in the US. When one looks at ES strategies outside of the United States it becomes clear that the ESD is less dominant in regions such as Europe and East Asia.[169] The conventional nature of the ESD is becoming increasingly challenged by ES discourses which place more value on renewable ES, and this is best illustrated by the way that renewable strategies are becoming intrinsic to East Asia's "new developmentalism" which has emerged to respond to various regional ES challenges.[170] Thus it must be stressed that while it is prevalent within the US context (within which other discourses also exist at the margins), the ESD only represents one of many other ES discourses.

Although it can accommodate the liberal political economy tradition of acquisition, and Chapter 4 will outline how it does so in relation to the trend of M&A which defined relations between Western oil companies at the turn of the century, the ESD reads energy acquisition as ultimately dependent upon neo-mercantilist approaches. Chapter 4 will illustrate how the ESD, utilized by US elites, discarded free market principles in favour of neo-mercantilist means when it became clear that CNOOC was using market mechanisms in its attempt to acquire Unocal.

Thus, although the ESD can operate in an environment in which positive security values are mobilized by the ES approach, the ESD perceives ES to ultimately be defined in negative security terms. Therefore, although the ESD can

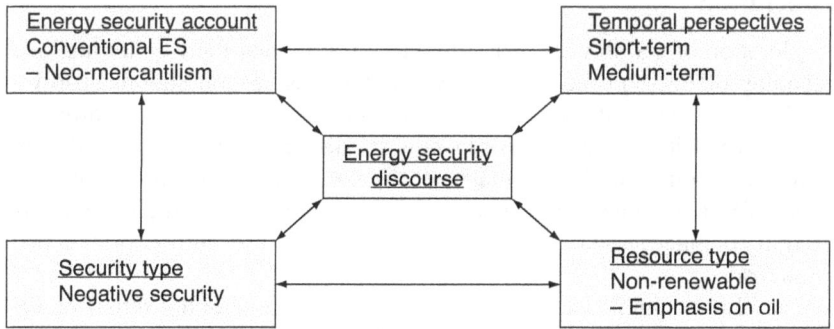

Figure 3.3 The Energy Security Discourse construct.

extend some way above the *x*-axis into the P1 quadrant, when placed under pressure, it tends to retreat into the N1 quadrant of Figure 3.1, and this is what was demonstrated by the CNOOC/Unocal affair. Similarly, while the ESD does account for medium-term ES concerns, it is primarily driven by interests dictated by short-term strategies as it focuses primarily on the affordability and accessibility of resources with a secondary concern for the availability of resources. While the ESD acknowledges development in techniques which prolong sustainability, such as deeper offshore drilling and fracking, it does not account for longer-term acquisition techniques which aim to harness renewable energies.

Thus, the ESD focus remains solely on the acquisition of non-renewable resources. As well, the examinations of official and non-official readings of ES illustrate that within the non-renewable resource realm, the ESD privileges oil as the non-renewable resource essential to ES policy. Although other non-renewables, such as coal and gas, are considered by the ESD, this occurs when scarcity of oil means that its affordability or accessibility warrants broader exploration into non-renewable alternatives. While the ESD can address notions of positive gains with regard to cooperation in accessing increasingly inaccessible resources (quadrant P1), it reads ES as a struggle for relative-gains as resources are ultimately non-renewable.

While no actor has explicitly referenced the ESD, I have argued that a continuity of certain policy objectives (i.e. availability, accessibility, and affordability) within particular ES strategies provide evidence enough to support the conclusion that this particular discourse has been mobilized as *the* approach to ES by the US. Therefore, while there is no direct textual mention of the ESD by any actor the actions of US elites and audience, and their inferences to universal epistemic assumptions and structures of ES, suggests that a particular ES discourse has been mobilized and is functioning at high levels of policy within the United States. This discourse is the ESD, and this chapter has sought to demonstrate how poststructural discourse analysis can be used to uncover the discursive constructions which advocate the existence of this particular, cohesive, and highly influential approach.

3.5 Conclusion

In its exploration of ES, this chapter aimed to illustrate that there is not just one understanding of ES, and that each understanding has several significant divisions within it. Because the ESD is but one understanding of ES, in order to examine the ESD it is essential to see how its discursive construction differs from other ES accounts so that we may see why it has gained the popular acceptance it has. ES is a contestable concept, and this can be highlighted when it is examined in conjunction with a case study because the case study can influence how ES is read.

There is also a significant divide between conventional and alternative ES accounts. Conventional accounts are more predominant than alternative accounts meaning that alternative ES readings tend to populate the margins of ES policy, although they are articulated more loudly within some ES circles. Conventional ES accounts emphasize neo-mercantilist and realist strategies for acquiring their energy. Such emphasis does not insinuate that proponents of conventional ES will not first utilize liberal market economy strategies to attain their energy requirements, but rather that these proponents will adopt neo-mercantilist strategies if the market fails to supply the resources they require. In this sense, the neo-mercantilist strategy, while not always employed, always exists as insurance. Thus, conventional ES accounts also privilege zero-sum benefits and relative gains, the ultimate goal of which is energy independence. These aspects of ES are directly linked to the prioritization of non-renewable sources within ES readings. Increasingly, these readings prioritize oil among other non-renewable resources, or in extremes, ES readings even equate oil with non-renewable resources with an emphasis on peak oil, scarcity, and the oil weapon.

The ESD places particular emphasis on oil, the acquisition of which is best achieved through neo-mercantilist strategies, if and when market strategies do not work. Therefore, when the ESD, which emphasizes relative gains, is twinned with the CTD within wider American discourse a situation emerges where China, and energy itself, become positioned as possible threats to US national security. Chapter 4 examines this in greater detail with its focus on the CNOOC/Unocal affair.

Notes

1 Dent, Christopher M. 2012. "Renewable Energy and East Asia's New Developmentalism: Towards a Low Carbon Future?" *The Pacific Review* 25(5): 561–587.
2 Dent, Christopher M. and Elspeth Thomson. 2013. "Asia's and Europe's Energy Policy Challenges: Introduction". *Asia Europe Journal* 11(3): 201–210.
3 Samuelson, Robert J. 2005. "China's Oil Bid: A Battle to Avoid…". *Washington Post*, 6 June.
4 Broomfield, Emma V. 2003. "Perceptions of Danger: The China Threat Theory". *Journal of Contemporary China* 12(35): 265–284.
5 Samaras, Constantine and Henry H. Willis. 2013. *Capabilities-Based Planning for Security at Department of Defense Installations*. Santa Monica, CA: Homeland Security and Defense Center (RAND).
6 Wendt, Alexander. 1987. "The Agent-Structure Problem in International Relations Theory". *International Organization*, 41(3): 335–370.
7 Environmental Protection Agency (EPA). 2013. "Clean Energy". *United States Environmental Protection Agency*. From www.epa.gov/cleanenergy/energy-and-you/affect/natural-gas.html, 25 September 2013.
8 Campbell, Colin J. 2006. "Understanding Peak Oil". [Online] *Peakoil.net (Association for the Study of Peak Oil & Gas)*, Retrieved 2 August 2013, from www.peakoil.net/about-peak-oil.
9 Wendt, Alexander. 1999. *Social Theory of International Politics*. Cambridge: Cambridge University Press, p. 112.
10 Winzer, Christian. 2012. "Conceptualizing Energy Security". *Energy Policy* 46: 36–48.
11 Checchi, Arianna, Arno Behrens, and Christian Egenhofer. 2009. "Long-Term Energy Security Risks for Europe: A Sector-Specific Approach". *Centre for European Policy Studies* 309: 1–47.

12 Ciuta, Felix. 2010. "Conceptual Notes on Energy Security: Total or Banal Security?" *Security Dialogue* 41(123): 123–144, p. 124.

13 Dent and Thomson. 2013. "Asia's and Europe's Energy Policy Challenges", p. 202.

14 Bardi, Ugo. 2013. "The Grand Challenge of the Energy Transition". *Frontiers in Energy Research*, 1: 1–4, p. 1.

15 Yergin, Daniel. 1993. *The Prize: The Epic Quest for Oil, Money and Power*. London: Pocket Books, p. 559.

16 Ciuta. 2010. "Conceptual Notes on Energy Security", p. 124.

17 International Energy Agency (IEA). 2013. "Topic: Renewables". *International Energy Agency*. Retrieved 4 October 2013, from www.iea.org/topics/renewables/.

18 Mingyuan, Wang. 2005. "Government Incentives to Promote Renewable Energy in the United States". *Temple Journal of Science, Technology & Environmental Law* 24(1): 355–366, p. 356.

19 Mohitpour, M. 2008. *Energy Supply and Pipeline Transportation: Challenges & Opportunities*. New York, ASME, p. 196.

20 Sieminski, Adam E. 2005. "World Energy Futures". In *Energy and Security: Toward a New Foreign Policy Strategy*, edited by Jan H. Kalicki and David L. Goldwyn. Baltimore: The Johns Hopkins University Press, p. 45.

21 BP. 2014. "BP Energy Outlook 2035". *BP.com/energy outlook*, Retrieved 16 February 2014, from www.bp.com/content/dam/bp/pdf/Energy-economics/Energy-Outlook/Energy_Outlook_2035_booklet.pdf.

22 Andrews, Edmund L. 2007. "Candidates Offer Different Views on Energy Policy". *New York Times*, 28 November.

23 "Why is Renewable Energy So Expensive?" 2014. *The Economist*, 5 January.

24 BP. 2012. "The BP Energy Outlook 2030". *BP.com*. Retrieved 24 February 2012, from www.bp.com/genericarticle.do?categoryId=9003467&contentId=7067432.

25 BP. 2012. "The BP Energy Outlook 2030".

26 Dent. 2012. "Renewable Energy and East Asia's New Developmentalism", p. 562.

27 Dent. 2012. "Renewable Energy and East Asia's New Developmentalism", p. 571.

28 World Nuclear Association. 2015. "Renewable Energy and Electricity". *World Nuclear Association*, Retrieved 18 March 2015, from www.world-nuclear.org/info/Energy-and-Environment/Renewable-Energy-and-Electricity/.

29 IEA. 2013. *World Energy Outlook 2013*. Paris, OECD/IEA.

30 IEA. 2012. "Renewable Energy Outlook: A Shining Future?" *World Energy Outlook 2012*. Paris: OECD/IEA, pp. 211–241.

31 Jowsey. 2009. "Economic Aspects of Natural Resource Exploitation", p. 303.

32 Angelis-Dimakis, Athanasios, Marcus Biberacher, Javier Dominguez, Guilia Fiorese, Sabine Gadocha, Edgard Gnansounou, Giorgio Guariso, Avraam Kartaldis, Luis Panichelli, Irene Pinedo, and Michella Roba. 2011. "Methods and Tools to Evaluate the Availability of Renewable Energy Sources". *Renewable and Sustainable Energy Reviews* 15: 1182–1200.

33 Bardi. 2013. "The Grand Challenge of the Energy Transition".

34 Jevons, W. Stanley. 1865. *The Coal Question: An Inquest Concerning the Progress of the Nation, and the Probable Exhaustion of Our Coal-Mines*. London and Cambridge: Macmillan and Co., p. viii.

35 IEA. 2013. *2013 Key World Energy Statistics*. Paris: OECD/IEA, p. 7.

36 IEA. 2013. *2013 Key World Energy Statistics*, p. 7.

37 IEA. 2013. *Resources to Reserves 2013: Oil, Gas and Coal Technologies for the Energy Markets of the Future*. Paris: OECD/IEA.

38 IEA. 2013. *Resources to Reserves 2013*.

39 IEA. 2013. *Resources to Reserves 2013*, p. 18.

40 Ciuta, Felix. 2009. "Security and the Problem of Context: A Hermeneutical Critique of Securitisation Theory". *Review of International Studies* 35: 301–326, p. 301.

41 Buzan, Barry. 1991. *People, States, and Fear: The National Security Problem in International Relations*. London: Harvester Wheatsheaf, pp. 18–19.
42 Buzan Barry and Ole Waever. 2003. *Regions and Powers: The Structure of International Security*. Cambridge: Cambridge University Press, p. 48.
43 Ciuta. 2009. "Security and the Problem of Context", p. 303.
44 Gjorv, Gunhild Hoogensen. 2012. "Security by Any Other Name: Negative Security, Positive Security, and a Multi-Actor Security Approach". *Review of International Studies* 38(4): 835–859, p. 837.
45 Gjorv. 2012. "Security by Any Other Name", p. 836.
46 Wolfers, Arnold. 1962. *Discord and Collaboration: Essays on International Politics*. Baltimore: Johns Hopkins Press, p. 153.
47 Gjorv. 2012. "Security by Any Other Name", p. 836.
48 Winzer. 2012. "Conceptualizing Energy Security", p. 38.
49 Huysmans, Jef. 2011. "What's in an Act? On Security Speech Acts and Little Security Nothings". *Security Dialogue* 42: 371–383, p. 373.
50 Obama, Barak. 2009. "Remarks by the President on Jobs, Energy Independence, and Climate Change". In *The American Presidency Project*, edited by Gerhard Peters and John T. Woolley. Retrieved 22 September 2014, from www.presidency.ucsb.edu/ws/index.php?pid=85689&st=&st1=, 4 October 2009.
51 Gjorv. 2012. "Security by Any Other Name", p. 836.
52 Booth, Ken. 1991. "Security and Emancipation". *Review of International Studies* 17(4): 313–326, p. 319.
53 Downs, Erica. 2004. "The Chinese Energy Security Debate". *The China Quarterly* 177: 21–41, p. 21.
54 Asia Pacific Energy Research Centre (APERC). 2007. *A Quest for Energy Security in the 21st Century*. Japan: Institute of Energy Economics.
55 APERC. 2007. *A Quest for Energy Security in the 21st Century*.
56 Kruyt, Bert, D.P. van Vuuren, H.J.M. de Vries, and H. Groenenberg. 2009. "Indicators for Energy Security". *Energy Policy* 37: 2166–2181.
57 Humphreys, Jasper. 2012. "Resource Wars: Searching for a New Definition". *International Affairs* 88(5): 1065–1082, p. 1065.
58 Loschel, Andreas, Ulf Moslener, and Dirk T.G. Rubbelke. 2010. "Indicators of Energy Security in Industrialised Countries". *Energy Policy* 38: 1665–1671.
59 BP. 2014. "Review By Energy Type: Hydroelectricity". *BP.com*. Retrieved 5 October 2013, from www.bp.com/en/global/corporate/about-bp/energy-economics/statistical-review-of-world-energy-2013/review-by-energy-type/hydroelectricity.html.
60 APERC. 2007. *A Quest for Energy Security in the 21st Century*.
61 IEA. 2013. *World Energy Outlook 2013*. Paris, OECD/IEA.
62 Smith, James L. 2012. "On the Portents of Peak Oil (And Other Indicators of Resource Scarcity)". *Energy Policy* 44: 68–78, p. 68.
63 Brandt, Adam R. 2007. "Testing Hubbert". *Energy Policy* 35(2007): 3074–3088, p. 3075.
64 Campbell, Colin J. 1988. *The Coming Oil Crisis*, Multi-Science Publishing Company & Petroconsultants S.A., p. 67.
65 Yergin. 1993. *The Prize: The Epic Quest for Oil, Money and Power*.
66 Galbraith, Kate. 2013. "Deep-Sea Drilling Muddies Political Waters". *New York Times*, 6 February.
67 Hargreaves, Steve. 2011. "Oil's Future in Deepwater Drilling". *CNN Money*. From http://money.cnn.com/2011/01/11/news/economy/oil_drilling_deepwater/, 11 January 2011.
68 Smith, Karl. 2013. "Will Natural Gas Stay Cheap Enough to Replace Coal and Lower US Carbon Emissions". *Forbes*, 22 March.
69 Kalicki, Jan H. and David L. Goldwyn. 2005. *Energy and Security: Toward a New Foreign Policy Strategy*. Baltimore: The Johns Hopkins University Press, p. 2.

70 Fagan, Mary. 2000. "Sheikh Yamani Predicts Price Crash as Age of Oil Ends". *The Telegraph*, 25 June.
71 Fagiani, Riccardo, Julian Barquin, and Rudi Hakvoort. 2013. "Risk-Based Assessment of the Cost-Efficiency and the Effectivity of Renewable Energy Support Schemes: Certificate Markets Versus Feed-In Tariffs". *Energy Policy* 55: 648–661, p. 650.
72 Paish, Oliver. 2002. "Small Hydro Power: Technology and Current Status". *Renewable and Sustainable Energy Reviews* 6: 537–556, p. 548.
73 Helm, Dieter. 2011. "Peak Oil and Energy Price – A Critique". *Oxford Review of Economic Policy* 27(1): 68–91.
74 Dannreuther, Roland. 2010. "Energy Security". In *The Routledge Handbook of New Security Studies*, edited by J. Peter Burgess. London: Routledge: 144–154, p. 151.
75 Regnier, Eva. 2007. "Oil and Energy Price Volatility". *Energy Economics* 29: 405–427.
76 BP. 2011. "Statistical Review of World Energy 2011". *BP.com*. Retrieved 24 February 2012, from www.bp.com/sectionbodycopy.do?categoryId=7500&contentId=7068481.
77 Narodoslawsky, Michael, Anneliese Niederl-Schmidinger, and Laszlo Halasz. 2008. "Utilizing Renewable Resources Economically: New Challenges and Chances for Process Development". *Journal of Cleaner Production* 16: 164–170, p. 165.
78 APERC. 2007. *A Quest for Energy Security in the 21st Century*.
79 IEA. 2013. "Energy Efficiency". *International Energy Agency*. Retrieved 4 October 2013, from www.iea.org/topics/energyefficiency/.
80 Martin, W.F. and E.M. Harrje, "The International Energy Agency", in *Energy and Security: Toward a New Foreign Policy Strategy*, edited by J.H. Kalicki., D.L. Goldwyn. Woodrow Wilson Press, Washington, p. 98.
81 Yergin, Daniel. 2005. "Energy Security and Markets". In *Energy and Security: Toward a New Foreign Policy Strategy*, edited by Jan H. Kalicki and David L. Goldwyn. Baltimore: The Johns Hopkins University Press, p. 51.
82 Dannreuther. 2010. "Energy Security", pp. 144–154.
83 Jowsey, Ernie. "Economic Aspects of Natural Resource Exploitation", *International Journal of Sustainable Development & World Ecology* 16(5): 303–307, p. 306.
84 Joshi, Ketan. 2014. "Why that Guy You Know Hates Renewable Energy". *Limited News*. From http://limitednews.com.au/2014/02/why-that-guy-you-know-hates-renewable-energy/, 13 February 2014.
85 Sklar, Scott. 2012. "American Exceptionalism and Renewable Energy: What the Tea Party Missed in 2011". *RenewableEnergyWorld.com*, from www.renewableenergyworld.com/rea/news/article/2012/01/American-exceptionalism-and-renewable-energy-what-the-tea-party-missed-in-2011, 4 January 2012.
86 Wustenhagen, Rolf, Maarten Wolsink, and Mary Jean Burer. 2007. "Social Acceptance of Renewable Energy Innovation: An Introduction to the Concept". *Energy Policy* 35: 2683–2691, p. 2683.
87 Tippee, Bob. 2012 "Defining Energy Security". *Oil & Gas Journal*. 23 January.
88 Brown, Matthew H., Christie Rewey, and Troy Gagliano. 2003. *Energy Security*. Boulder, CO: National Conference of State Legislatures, p. 1.
89 Jordan, Matt, Dawn Manley, Valerie Peters, and Ron Stoltz. 2012. *The Goals of Energy Policy: Professional Perspectives on Energy Security, Economics, and the Environment*. Albuquerque, NM: Sandia National Laboratories/U.S. Department of Energy, p. 2.
90 Kerr, David. 1999. "The Chinese and Russian Energy Sectors: Comparative Change and Potential Interaction". *Post-Communist Economies* 11(3): 337–372, pp. 337–338.
91 Abraham, Spencer. 2001. "Meet Energy Challenges; Start Providing Security at Home". *Washington Times*, 29 November.

92 Nixon, Richard. 1973. "Remarks About the Nation's Energy Policy". [Online] In *The American Presidency Project*, edited by Gerhard Peters and John T. Woolley. From www.presidency.ucsb.edu/ws/index.php?pid=3953&st=energy&st1=oil, 8 September 1973.

93 Ford, Gerald R. 1977. "Annual Message to the Congress: The Economic Report of the President". [Online] In *The American Presidency Project*, edited by Gerhard Peters and John T. Woolley. From www.presidency.ucsb.edu/ws/index.php?pid=557 0&st=oil&st1=energy+security, 18 January 1977.

94 Nixon, Richard. 1973. "Address to the Nation About Policies to Deal With the Energy Shortages". [Online] In *The American Presidency Project*, edited by Gerhard Peters and John T. Woolley. From www.presidency.ucsb.edu/ws/index.php?pid=403 4&st=project+independence&st1=, 7 November 1973.

95 Carter, Jimmy. 1979. "Kansas City, Missouri Remarks at a Reception for Business and Civic Leaders". [Online] In *The American Presidency Project*, edited by Gerhard Peters and John T. Woolley. From www.presidency.ucsb.edu/ws/index.php ?pid=31533&st=cheap+oil&st1=, 15 October 1979.

96 "How Gas Price Controls Sparked '70s Shortages". 2006. *Washington Times*, 15 May.

97 Yergin. 1993. *The Prize: The Epic Quest for Oil, Money and Power.*

98 Reagan, Ronald. 1982. "Remarks at the Opening Ceremonies for the Knoxville International Energy Exposition (World's Fair) in Tennessee". [Online] In *The American Presidency Project*, edited by Gerhard Peters and John T. Woolley. From www.presidency.ucsb.edu/ws/index.php?pid=42470&st=Remarks+at+the+Opening +Ceremonies+for+the+Knoxville+International+Energy+Exposition+%28World%5 C%27s+Fair%29+in+Tennessee&st1=.

99 Mufson, Steven. 2008. "Oil Closes Over $100 for 1st Time". *Washington Post*, 20 February.

100 Bush, George. 1990. "The President's News Conference on the Persian Gulf Crisis". [Online] In *The American Presidency Project*, edited by Gerhard Peters and John T. Woolley. From www.presidency.ucsb.edu/ws/index.php?pid=18792&st=oil&st1= iraq, 30 August 1990.

101 Clinton, William J. 1995. "Statement on Petroleum Imports and Energy Security". [Online] In *The American Presidency Project*, edited by Gerhard Peters and John T. Woolley. From www.presidency.ucsb.edu/ws/index.php?pid=50988&st=energy+sec urity&st1=oil, 16 February 1995.

102 Clinton, William J. 2000. "Statement on the Organization of Petroleum Exporting Countries Production Decision and the Legislative Agenda for Energy Security". [Online] In *The American Presidency Project*, edited by Gerhard Peters and John T. Woolley. From www.presidency.ucsb.edu/ws/index.php?pid=58301&st=energy+sec urity&st1=oil, 28 March 2000.

103 Bush, George W. 2001. "Remarks Announcing the Energy Plan in St. Paul, Minne-sota". [Online] In *The American Presidency Project*, edited by Gerhard Peters and John T. Woolley. From www.presidency.ucsb.edu/ws/index.php?pid=45617&st=ene rgy+security&st1=oil, 17 May 2001.

104 Friedman, Thomas L. 2003. "A War for Oil?" *New York Times*, 5 January.

105 Zichal, Heather. 2012. "Increasing Energy Security". [Online] *U.S. Department of Energy*, from http://energy.gov/articles/increasing-energy-security.

106 Nixon, Richard. 1973. "Remarks on Transmitting a Special Message to the Congress on Energy Policy". [Online] In *The American Presidency Project*, edited by Gerhard Peters and John T. Woolley. From www.presidency.ucsb.edu/ws/index.php?pid=381 6&st=outer+continental+shelf&st1=, 18 April 1973.

107 Carter, Jimmy. 1977. "Address to the Nation on Energy". [Online] In *The American Presidency Project*, edited by Gerhard Peters and John T. Woolley. From www.pres-idency.ucsb.edu/ws/index.php?pid=7369&st=reduce+demand+through+conservation &st1=.

108   Washington Institute. 1991. [Online] *The Washington Institute for Near East Policy*, from www.washingtoninstitute.org/policy-analysis/view/a-postgulf-war-assessment.
109   Brookings Institution. 2014. "About the Energy Security Initiative". [Online] *Energy Security Initiative*. From www.brookings.edu/about/projects/energy-security/about.
110   Brookings Institution. 2014. "About the Energy Security Initiative".
111   Jordan *et al.* 2012. *The Goals of Energy Policy*, pp. 2–4.
112   Jordan *et al.* 2012. *The Goals of Energy Policy*, p. 2.
113   Tippee. 2012. "Defining Energy Security".
114   Tippee. 2012. "Defining Energy Security".
115   Brown *et al.* 2003. *Energy Security*, pp. 9–11.
116   Brown *et al.* 2003. *Energy Security*, p. 8.
117   Kruyt *et al.* 2009. "Indicators for Energy Security".
118   Sharan, Sunil. 2011. "America is Losing the Green Energy Race". *Washington Post*, 7 December.
119   Arora-Jonsson, Seema. 2011. "Virtue and Vulnerability: Discourses on Women, Gender and Climate Change". *Global Environmental Change* 21: 744–751, p. 744.
120   Resurrección, Bernadette P. 2013. "Persistent Women and Environmental Linkages in Climate Change and Sustainable Development Agendas". *Women's Studies International Forum* 40(2013), 33–43, p. 33.
121   Dalby, Simon. 2002. *Environmental Security*. Minneapolis: University of Minnesota Press, p. 126.
122   Alston, Margaret. 2013. "Gender Mainstreaming and Climate Change". *Women's Studies International Forum* 47: 287–294, p. 288.
123   Brady, David and Denise Kall. 2008. "Nearly Universal, but Somewhat Distinct: The Feminization of Poverty in Affluent Western Democracies, 1969–2000". *Social Science Research* 37: 976–1007.
124   Tobak, Steve. 2013. "Don't Buy an Electric Car". *FOX Business*. Retrieved 5 November 2013, from www.foxbusiness.com/business-leaders/2013/04/05/dont-buy-electric-car/?intcmp=sem_outloud.
125   Bryce, Robert. 2010. "Five Myths About Green Energy". *Washington Post*, 25 April.
126   Moms, Gijs. 2004. *Electric Vehicle: Technology and Expectations in the Automobile Age*. Baltimore: Johns Hopkins University Press.
127   Bryant Jr., Keith L. 1988. *Encyclopedia of American Business History and Biography: Railroads in the Age of Regulation, 1900–1980*. New York: Facts on File.
128   Moms. 2004. *Electric Vehicle*.
129   Scharff, Virginia. 1991. *Taking the Wheel: Women and the Coming of the Motor Age*. New York: Free Press.
130   Winter, Nicholas J.G. 2010. "Masculine Republicans and Feminine Democrats: Gender and Americans' Explicit and Implicit Images of the Political Parties". *Political Behavior* 34(2): 587–618, p. 591.
131   Newport, Frank. 2009. "Women More Likely to Be Democrats, Regardless of Age". *Gallup*, from www.gallup.com/poll/120839/women-likely-democrats-regardless-age.aspx, 12 June 2009.
132   Langer, Gary. 2013. "Poll Finds Vast Gaps in Basic Views on Gender, Race, Religion and Politics". *ABC News*. From http://abcnews.go.com/blogs/politics/2013/10/polll-finds-vast-gaps-in-basic-views-on-gender-race-religion-and-politics/, 28 October 2013.
133   Red State. 2008. "Democrats and Sissies". *Red State*. Retrieved 5 October 2013, from http://archive.redstate.com/stories/the_parties/democrats/democrats_and_sissies.
134   Hoft, Jim. 2013. "Study: Weak Sissy Men More Likely to Support Welfare State, Wealth Redistribution, Democrats". *Gateway Pundit*. Retrieved 5 October 2013, from www.thegatewaypundit.com/2013/05/study-weak-sissy-men-more-likely-to-support-welfare-state-wealth-redistribution-democrats/.
135   Winter. 2010. "Masculine Republicans and Feminine Democrats", p. 588.

136 Lane, Charles. 2012. "Liberals' Green-Energy Contradictions". *Washington Post*, 15 October.
137 Cappiello, Dina and Matthew Daly. 2012. "Republicans, Democrats at Odds on Energy Issues". *The Associated Press-NORC Center for Public Affairs Research*. From www.apnorc.org/news-media/Pages/News+Media/republicans-democrats-at-odds-on-energy-issues.aspx, 14 June 2012.
138 Sovacool, Benjamin K. 2009. "The Importance of Comprehensiveness in Renewable Electricity and Energy Efficiency Policy". *Energy Policy* 37: 1529–1541, p. 1530.
139 Brookings Institution. 2014. "About the Energy Security Initiative".
140 Kruyt *et al.* 2009. "Indicators for Energy Security", p. 2166.
141 Kruyt *et al.* 2009. "Indicators for Energy Security", p. 2167.
142 Ciarreta, Aitor, Maria Paz Espinosa, and Cristina Pizarro-Irizar. 2014. "Is Green Energy Expensive? Empirical Evidence from the Spanish Electricity Market". *Energy Policy* 69: 205–215, p. 215.
143 Intergovernmental Panel on Climate Change (IPCC). 2000. *IPCC Special Report: Emissions Scenarios – Summary for Policymakers*. Intergovernmental Panel on Climate Change, Intergovernmental Panel on Climate Change: 1–21, p. 4.
144 IPCC. 2000. *IPCC Special Report*, p. 4.
145 Schwartz, Nelson D. 2004. "Inside the Head of BP" *Fortune Magazine*, 26 July.
146 Najam, Adil, David Runnalls, and Mark Halle. 2007. *Environment and Globalization: Five Propositions*. Winnipeg, Manitoba: International Institute for Sustainable Development.
147 International Renewable Energy Agency (IRENA). 2014. "About IRENA". *International Renewable Energy Agency (IRENA)*. From www.irena.org/Menu/index.aspx?PriMenuID=13&mnu=Pri.
148 Nixon, Richard. 1973. "Address to the Nation About National Energy Policy". In *The American Presidency Project*, edited by Gerhard Peters and John T. Woolley. From www.presidency.ucsb.edu/ws/?pid=4051, 25 November 1973.
149 Carter, Jimmy. 1977. "The Environment Message to the Congress". In *The American Presidency Project*, edited by Gerhard Peters and John T. Woolley. From www.presidency.ucsb.edu/ws/index.php?pid=7561&st=As+our+nation+increasingly+turns+to+coal+as+a+replacement+for+our+dwindling+supplies+of+oil+and+gas%2C+we+must+be+sure+that+we+will+not+fall+short+of+the+goals+we+have+established+to+protect+human+health+and+the+general+environment&st1=.
150 Kruyt *et al.* 2009. "Indicators for Energy Security", p. 2167.
151 Yergin. 1993. *The Prize: The Epic Quest for Oil, Money and Power*, p. 111.
152 Checchi *et al.* 2009. "Long-Term Energy Security Risks for Europe", p. 1.
153 Singer, Clifford. 2008. "Oil and Security". *Policy Analysis Brief*. Muscatine, IA: The Stanley Foundation, p. 1.
154 Dannreuther. 2010. "Energy Security", p. 148.
155 Mufson, Steven. 2013. "Does OPEC Still Have the U.S. Over a Barrel?" *Washington Post*, 11 October.
156 Downs, Erica. 2010. "Who's Afraid of China's National Oil Companies?" In *Energy Security: Economics, Politics, Strategies and Implications*, edited by Carlos Pascual and Jonathan Elkind. Washington, DC: Brookings.
157 Mufson. 2013. "Does OPEC Still Have the U.S. Over a Barrel?"
158 Bush, George W. 2008. "Remarks on Energy and Climate Change". In *The American Presidency Project*, edited by Gerhard Peters and John T. Woolley. From www.presidency.ucsb.edu/ws/index.php?pid=76957&st=international+cooperation+and+technology+investment&st1=.
159 Dannreuther. 2010. "Energy Security", p. 152.
160 Dannreuther. 2010. "Energy Security", p. 153.
161 Eilperin, Juliet. 2010. "Emissions Limits, Greater Fuel Efficiency for Cars, Light Trucks Made Official". *Washington Post*, 2 April.

162 Whoriskey, Peter. 2011. "Conventional Gas-Powered Cars Starting to Match Hybrids in Fuel Efficiency". *Washington Post*, 9 March.
163 OCHA. 2010. *Energy Security and Humanitarian Action: Key Emerging Trends and Challenges*. OCHA: Policy Development and Studies Branch, UN Office for the Coordination of Humanitarian Affairs (OCHA): 1–14.
164 Linklater, Andrew. 2005. "Political Community and Human Security". In *Critical Security Studies and World Politics*, edited by Ken Booth. Boulder: Colorado, Lynne Rienner Publishers, Inc., 113–133, p. 124.
165 Dalby. 2002. *Environmental Security*, p. 7.
166 Dannreuther. 2010. "Energy Security", p. 147.
167 Sieminski. 2005. "World Energy Futures", p. 21.
168 Dannreuther. 2010. "Energy Security", p. 153.
169 Dent. 2012. "Renewable Energy and East Asia's New Developmentalism".
170 Dent. 2012. "Renewable Energy and East Asia's New Developmentalism".

4 Case study
The CNOOC/Unocal affair

4.1 Introduction

In the summer of 2005, amidst much controversy, CNOOC made what was an eventually failed bid to purchase the American oil company Unocal. The controversy which surrounded the deal was a result of the way in which the China Threat Discourse interacted with the Energy Security Discourse in broader American discourse whereby conceptions of energy security became defined, in part, by the CTD. Oil had already become a strategic commodity central to countries' national interests and subsequent foreign policies, but I argue that the CTD magnified already sensitive elements of the ESD. The failure of CNOOC's bid for Unocal suggested that the China threat was so pervasive in American discourse that an otherwise legal transaction could be scuppered due to the perception of China as a threat to US interests. The decade from the mid-1990s to the mid-2000s was one in which foreign M&A and corporate consolidation defined the oil industry and examples of successful foreign acquisitions of American oil companies are provided in section 4.2 to demonstrate the exceptional nature of the CNOOC/Unocal affair.

CNOOC faced pressure from the American establishment to withdraw its bid and a CNOOC statement expressed its disappointment saying that "the unprecedented political opposition ... was regrettable and unjustified".[1] Evidence suggests that while the ESD provided the central discourse within which ES was read in the US, sales of American oil companies to foreign bodies were not perceived as threats to US national security in and of themselves. This was, in fact, actually somewhat common at the turn of the century. What was exceptional was the way in which a bid for Unocal by the Chinese state-owned CNOOC was immediately positioned as a threat. This case was unique because the CTD played a specific role in amplifying the security aspect of the ESD. The CTD and the ESD became so entwined with respect to the CNOOC/Unocal deal that congressman Joe Barton (R-TX) argued that the deal should be disallowed, stating that "this transaction poses a clear threat to the energy and national security of the United States".[2]

Admittedly, American opposition to the sale was not universal and there were some who argued that political interference into such an interaction was itself harmful to US interests as it was contrary to the American ideal of the free market. However, such arguments were marginalized and the CTD worked with the ESD to create a potent atmosphere of distrust towards CNOOC. Not unexpectedly, the American reaction was derided by the Chinese. Referring to the backlash which CNOOC's bid generated, Zhang Guobao, the vice chairman of China's National Development and Reform Commission, was quoted at the time as saying that "To spread the 'China Threat' and try to curb China's progress and starve its energy needs is not in the interest of world stability and development. Such attempts are doomed to fail".[3] However,

> American views of China have a documented history of oscillation between fear, affection, and strategic rivalry. In 2001, the incoming George W. Bush administration was quick to distance itself from the Clinton administration's recognition, however tepid and brief it was, of China as a strategic partner. Rather, China came to be viewed increasingly through the lens of strategic competition.[4]

The CNOOC/Unocal affair exemplifies the articulation of the CTD and ESD and there has been no case since which has highlighted the relationship between them with anywhere near as much clarity. Where Chapters 2 and 3 were devoted to the examinations of the CTD and the ESD in isolation, the goal of this chapter is to bring them together in an analysis of the CNOOC/Unocal affair. By examining the way CNOOC's bid for Unocal unravelled and eventually failed it will be demonstrated how the CTD and the ESD worked in concert to utilize fears of oil scarcity to elevate perceptions of China's threat to US national interests.

4.2 The central discourses in the CNOOC/Unocal debate

In order to demonstrate the importance of the case study it is good to return to the central discourses of the China Threat and Energy Security to examine how they worked in conjunction to create an environment in which the events of the CNOOC/Unocal affair transpired. I will illustrate how their joint deployment specifically impacts upon the CNOOC/Unocal case study in order to set the stage for the textual analysis of the official and non-official discourses which will follow.

In order to illustrate how the central discourses interacted in this case study it is helpful to first turn to Lene Hansen whose work helped provide the framework for this book. Hansen used discourse analysis to examine Western perceptions of the Bosnian war. In her 2006 book *Security as Practice* she demonstrates the way the Genocide Discourse superseded the Balkan Discourse in the Western debate on Bosnia when she writes that,

> The stability of the Balkan discourse is built on a firm articulation of spatial and temporal difference between 'the Balkans' and 'the West,' but this stability can only be upheld as long as the representation of the warfare as 'genocide' can be avoided.[5]

While the re-articulation of conflict in the Balkans as 'genocide' warranted the adoption of a new discourse, where the Balkan Discourse was superseded by the Genocide Discourse, the relationship between the CTD and the ESD is much different in that rather than displacing it, the latter actually helps to uphold the former. The ethical obligations the United States has towards China do not change with the advent of the ESD, but instead they augment those obligations, and the US is forced to further protect its own interests and 'domestic respons- ibilities'. In essence, the ESD helps to *legitimize* the CTD as it provides a lens through which ES data can be used to reinforce more ethereal notions of China's menace to the United States. There is no paradigm shift which occurs, but rather a paradigm entrenchment.

I will begin by examining the focus on oil by the American establishment in the years which preceded and followed CNOOC's 2005 Unocal bid in order to reaffirm the importance of the ESD and demonstrate the importance of the case study. Having demonstrated the central role that security of oil supply played in foreign policy considerations of the US we can then understand how the CTD amplified security concerns within the ESD with respect to CNOOC's Unocal bid. This examination will make clear how inexorably linked the CNOOC/ Unocal affair is to the central discourses of the China threat and energy security within wider American discourse and how it represents the most potent embodi- ment of this discursive relationship in practice.

4.2.1 The ESD and the CNOOC/Unocal affair

As discussed in Chapter 3, ES can be read in several ways. Although the CNOOC/Unocal affair suggests that the US response to the Chinese company's bid was overwhelmingly informed by the ESD, other readings did emerge. For instance, David L. Goldwyn, the former assistant secretary of energy during the Clinton administration, thought that the backlash against CNOOC was harmful to the US and wrote, "What this misguided policy did was to say the United States will not advocate fair trade when it comes to American assets.... That may push China to a more competitive stance rather than a more cooperative one".[6] However, despite such alternative readings Chapter 3 also argued that non-renewable resources are central to conceptions of the ESD which is itself essential to contemporary US understandings of ES. Non-ESD readings are mar- ginalized by the attention given to short-term negative security concerns of non- renewable resource acquisition. Had alternative ES discourses not been marginalized, the CNOOC/Unocal affair, concerned as it is with oil companies, would not have been as illustrative or important as I claim it to be, and for this reason I will demonstrate how non-renewable resources were prioritized by the US establishment at the expense of renewables in recent decades. Because I argue that the CNOOC/Unocal affair is both a product of, as well as a contrib- utor to the central discourses, this subsection will briefly examine the role renew- able and non-renewable resources played in US discourse prior to and following the events of the case study.

Although renewable energy implementation is often discussed as a matter of urgency within the American political establishment, actual implementation faces significant obstacles to success. Rep. Tom Udall (D-MN) argued that "To eliminate our dependence on foreign oil and develop a new economy based on clean, renewable non-polluting energy, we need a massive long-term investment".[7] However, E. Donald Elliott writes that "Shifting policies and changing priorities as different parties come to power in the United States has been one of the major difficulties that we have had in promoting renewable energy".[8] Thus, a major barrier to renewable energy implementation in the US is the lack of sustained policies between administrations. To illustrate the impact this has, Elliott explains that "The Germans have made a long-term commitment to buy renewable energy for 20 years, which facilitates developers in financing their projects. [America's] policies tend to come and go".[9] This inability to focus on long-term strategies helps to narrow focus on short-term strategies of non-renewable resource acquisition.

This short-term focus was highlighted in 2008 when a reporter asked President Bush why he hadn't put more resources into renewable energy development. The President responded that "The problem is, there's been a lot of focus by the Congress in the intermediate steps and in the long-term steps – the long-term steps being hydrogen; the intermediate steps being biofuels, for example, and researching the biofuels and battery technology – but not enough emphasis is on the here and now".[10] President Bush side-stepped the question by suggesting the lack of innovation resulted from Congressional inaction, but also suggested that long- and medium-term policies should be sacrificed for short-term gain. The focus on short-term ES strategies of non-renewable resource acquisition is due to the perception that the US energy infrastructure has become dependent upon it. Earlier in his Presidency Bush explained that

> We're in a transition period ... we're making changes as to how we use energy and how we supply energy.... But in the meantime, we're hooked. We import over 60 percent – or about 50-something percent of our energy supplies. And that means we've got to have a short-term energy policy that makes sense until technology changes.[11]

This short-term focus was much more in tune with the US emphasis on the ESD than the ES strategy suggested by the subsequent administration.

The temporal element (that is, short-term versus long-term concerns) of ES strategies is illustrative of problems which arise from shifting policies, and returning to an idea addressed in Chapter 3, these shifts often correlate with the party in power. Although both major US parties pay lip service to issues of renewable energy, Republicans tend to privilege short-term ES gains, and therefore non-renewable resources, more than Democrats.[A] So where President Bush

A This made the bipartisan response to CNOOC's Unocal bid all the more remarkable, especially considering the association between 'green' energy and Democrats, and Republican derision of long-term ES which was explored in Chapter 3.

argued that we need emphasis on the "here and now", President Obama, echoing President Carter, argued that there are "no quick fixes" to US ES, and "We're going to have to think long term".[12] President Obama's energy policy suggested a shift towards new energy technologies to curb pollution as well as dependence on foreign sources of energy. His policy strayed from the ESD as it favoured long-term implementation of renewables over short-term energy independence, and as it ventured away from the ESD it was met with derision by many in the US establishment who referred to it as a 'war on coal'.[B] Regarding President Obama's inaction on the Keystone pipeline and his 'war on coal' Congresswoman Martha Roby (R-AL) balked "A war on coal? A war on coal ultimately amounts to a war on American energy and a war on American families",[13] and Congressman Richard Hudson (R-NC) added "You know, we ought to have a war on gas prices".[14] This preoccupation with gas prices helps to situate the CNOOC/Unocal affair temporally within a political climate where, despite attempts at shifting policies at the executive level, adherence to the ESD remains strong. In a House Committee on Energy and Commerce meeting in 2001 Rep. Barton (R-TX) clarified the place of oil in US ES policy by stating "I drove here today in a car that uses gasoline. I think most of us probably arrived here by transportation that uses gasoline, also. The demand for petroleum is not going down".[15] Elliott suggests this is a result of the fact that "[American] history has deeply embedded the expectation of *cheap energy* in our citizens".[16] This expectation of cheap energy is represented by a culture of huge consumption in which the US has 40 percent dependence on net petroleum imports despite the fact that it is the world's fourth-largest producer.[17] Such an acute dependency on oil helps to ensure that long-term renewable gains are sacrificed for short-term non-renewable gains, and this ensures that the global oil market and companies retain their importance to US ES.

4.2.2 The impact of the CTD on the ESD: the CNOOC/Unocal affair as an exceptional case

The way in which the American establishment rallied to deny CNOOC its purchase of Unocal is indicative of the way the CTD has informed current perceptions of China by the US. From the late 1990s to the mid-2000s, the international oil market was one defined by "intense consolidation" and M&A (Table 4.1) as industry giants attempted to appease a market demanding growth by purchasing smaller rivals because "Acquisitions tend to be easier than organic methods".[18] This logic was sustained through several megamergers which involved foreign takeovers, international oil companies (IOCs), and American interests, and was only questioned when CNOOC made its unsolicited bid for Unocal in 2005.

It is important to briefly address some of the major foreign acquisitions during this period to demonstrate the exceptional nature of the CNOOC/Unocal affair and to show how the ESD mobilized on its own clearly differs from the ESD

B See: Roff (2013).

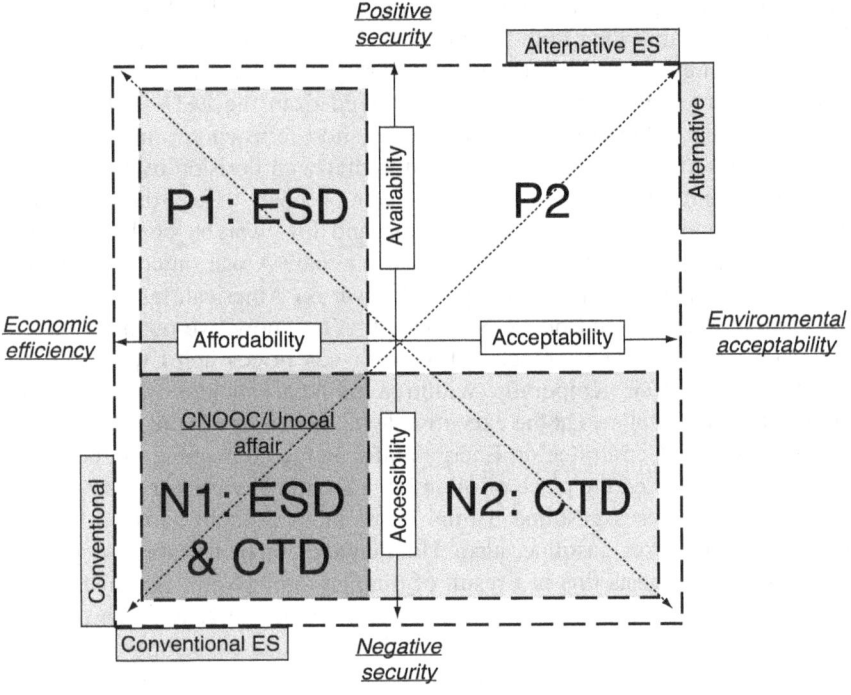

Figure 4.1 Intersection of the ESD and CTD.

Table 4.1 Major M&A during decade of consolidation

1997	Shell buys Texaco holdings in US Midwest and West Coast for $2 billion.
1997	Shell, Texaco, and Saudi Aramco merge East Coast and Gulf Coast operations for billions.
1998	BP buys Amoco for $48 billion.
1999	BP buys ARCO for $27 billion.
1999	TotalFina and Elf merge for €52.6 billion.
1999	Exxon buys Mobil for $82 billion.
2001	Chevron and Texaco merge operations for $39.5 billion.
2002	Phillips Petroleum Co and Conoco Inc. Merge for $18 billion.
2005	ConocoPhillips buys Burlington Resources Inc. for $65 billion.
2005	CNOOC bids $18.5 billion for Unocal Corp. FAILED.
2005	Chevron buys Unocal Corp. for $17.1 billion.

when it is mobilized in conjunction with the CTD. This is illustrated in Figure 4.1 where the ESD, representing all acquisitions not involving China, occupies both the positive and negative security to the left of the *y*-axis to encompass all issues of affordability regarding ES. This is represented by the lightly-shaded and dotted rectangle in P1 and N1.

The CNOOC/Unocal case is unique in that it is solely located in the negative security quadrant of N1 (represented by the darkly-shaded area), and its location results from the direct intervention of the CTD which emphasizes the negative security concerns of regionalization over the positive security of international cooperation which characterized the successful foreign acquisitions and mergers which did not involve Chinese companies.

Although small by standards which would be set in the years to follow, in 1997 two foreign interventions into the corporate structure of Texaco set the precedent that US oil companies were amenable to foreign investment. On 18 March 1997, Shell and Texaco announced a merger in their West Coast and Midwest refining and marketing operations in a venture which became known as Equilon Enterprises LLC.[19] In this instance, Shell, the foreign corporate partner, took 56 percent control of the venture, the assets of which surveyors estimated to be around £1 billion.[20] On 16 July of that year Texaco announced another foreign merger, this time with Saudi Aramco and Shell which would consolidate its East Coast and Gulf Coast refining operations wherein Shell would obtain 35 percent ownership, and Saudi Aramco and Texaco would each obtain 32.5 percent ownership.[21] These two ventures showed not only that US oil companies were open to foreign investment, but that they were also, as minority shareholders, open to foreign ownership.

However, the first real megamerger took place between BP and Amoco (formerly Standard Oil of Indiana) in 1998 when Britain's BP agreed to acquire Indiana's Amoco for $48 billion in a $110 billion merger "making it the biggest-ever industrial merger".[22] The merger created the largest US oil and gas producer and Britain's largest corporation (based in London), and it represents American ease regarding foreign adventures into its strategically vital energy industry. Moreover, in cost-saving measures, the companies agreed to lay off 6000 employees.[23,C] Upon news of the merger, shares in the company, which became 60 percent owned by BP and 40 percent owned by Amoco, surged.[24] Amoco had been the fourth-largest US oil producer[D] and its sale to BP created "the largest takeover of an American company by a foreign concern".[25] It is interesting to note that BP's acquisition was successful as Andrew Avramides, an oil consultant in London, explained that "The strength of Amoco is in areas that are lacking at B.P. such as the natural gas reserves which it possesses"[26]; the very same areas CNOOC was interested in Unocal for. Although the size of the acquisition warranted review by the Federal Trade Commission (FTC) in Washington, there were no obvious impediments or vocal opponents to it.[27]

BP again looked to American shores to consolidate its hold on the market when, in March of 1999, it announced its intent to buy California's ARCO (the Atlantic Richfield Company) for $27 billion, which it successfully acquired following "a year of negotiations with FTC regarding how the companies' merger

---

C   It will be demonstrated below that in the face of opposition based on fear of job cuts, CNOOC promised to refrain from cutting American jobs in its Unocal bid.

D   Amoco was far larger than Unocal.

would affect competition in Alaska and on the US West Coast".[28] The delay in the deal's finalization was due to uncertainty that BP's acquisition of ARCO would give it too much control of the Alaskan oil fields which remained an important element to US energy sovereignty.[29] As a long-time competitor, BP also faced criticism from ARCO's workforce who were shocked by the sale, with one Yahoo subscriber even stating that "All Arco employees will bow to the Queen of England and should have her portrait on their wall if they expect any chance in hell of remaining on the payroll after the merger".[30] Indeed, in an attempt to save $1 billion a year in costs,[31] BP was so vocal about its willingness to slash its American workforce that "BP Amoco Chief Executive Officer John Browne [was] likened to a neutron bomb for his cost-cutting zeal: Often called 'Neutron John,' he gets rid of the people while preserving the hard assets".[32] However, it is telling that although BP, as a foreign (yet Western) oil company, did face challenges, it was ultimately able to prove that its interests in ARCO, a US producer larger than Unocal, did not run counter to those of the US at the national level. The deal meant that BP would produce more oil than either Mobil-Exxon or Royal Dutch/Shell Group and of great importance was the fact that ARCO had "strong positions in natural gas in Indonesia and China, where BP Amoco wants to expand".[33] This was a position CNOOC was keen to emulate with its bid for Unocal.

These cases illustrate that rather than being insular and opposed to consolidation, US oil companies at the turn of the millennium were keen to explore foreign acquisitions in order to become more lean and profitable. Of greatest import is the fact that unlike the CNOOC/Unocal affair, economic considerations surrounding these deals were not impeded by overriding political concerns. The Shell-Texaco, Saudi Aramco-Texaco, BP-Amoco, and BP-ARCO deals serve as a few examples of success in the consolidation of oil companies on an international level, and US companies are central to each of them. If we look to examples of consolidation of Western oil companies more generally, even more potent examples emerge. These include the €52.6 billion Total SA merger between TotalFina and Elf in September 1999, the $82 billion purchase of Mobil by Exxon in November 1999, the $39.5 billion deal between Chevron and Texaco in September 2001, the $18 billion merger between Phillips Petroleum Co and Conoco Inc. in August 2002, and the £33.8 billion purchase of Burlington Resources Inc. by ConocoPhillips in December 2005.[34] All of these mergers were successful and fall within the ESD half of Figure 4.1, concerned as they all are with issues of availability, affordability, and accessibility. Because these mergers reflect emphasis on positive security gains for all parties through consolidation, the contrast between them and CNOOC's bid for Unocal becomes very stark as the CNOOC/Unocal affair is firmly located in the location of N1 due to the negative security approach the CTD engendered in the US establishment. In order to provide a base for further analysis the CNOOC/Unocal affair will now be explored in detail.

4.3 Overview: a timeline examination of CNOOC's bid for Unocal

This section provides a basic overview of the timeline of CNOOC's bid for Unocal and provides introductory evidence as to how the CTD interacted with the ESD. Table 4.2 provides a timeline of events.

On 4 April 2005, Chevron announced a bid to buyout its smaller American competitor Unocal for roughly $16.5 billion, and this was followed by antitrust approval for the acquisition which was given by the US Federal Trade Commission on 10 June of the same year.[35] However, this bid was trumped when, on 23 June, CNOOC made an offer for Unocal worth more than $18.5 billion. At this time Unocal was a relatively unremarkable mid-sized American oil company that many thought was not worth the value offered by the Chinese company. The fact that CNOOC offered such a large amount for Unocal was exceptional because of the Chinese company's relatively small size. CNOOC was not the largest oil company in China, but rather the third largest, after CNPC (PetroChina) and Sinopec (China Petrochemical Corporation). Moreover, CNOOC, like its Chinese NOC counterparts, was much smaller than the major IOCs, such as ExxonMobil, BP, or Royal Dutch Shell. To detractors of the sale, this detail led to concerns about two other factors, the first that the value of Unocal lay outside its monetary worth in the marketplace, and the second that CNOOC was working in concert with the Chinese government.

The fact that the CCP was instrumental to CNOOC's ability to make its bid for Unocal caused uproar in the US as it was argued that "CNOOC's Communist government ownership and promise of virtually interest-free loans are not

Table 4.2 Timeline of CNOOC's bid for Unocal

4 April 2005	Unocal is agreed to be sold to Chevron for US$16.5.
10 June 2005	Chevron granted antitrust approval by US Federal Trade Commission.
23 June 2005	CNOOC makes a bid for Unocal for US$18.5 billion.
30 June 2005	H.Res.344, introduced by Richard Pombo, is passed by the House. The resolution calls for a CFIUS review of CNOOC's bid.
30 June 2005	H.Amdt.431, introduced by Carolyn Kilpatrick, is amended to the appropriations bill H.R.3058 by the House. The amendment prohibits the use of Treasury funds to approve the sale of Unocal to CNOOC.
2 July 2005	CNOOC applies for CFIUS review.
20 July 2005	Chevron increases its offer for Unocal to $17.1 billion. The increased bid also increases the amount of cash offered by 40 percent.
2 August 2005	CNOOC withdraws its bid for Unocal.
8 August 2005	H.R.6/P.L.109–58 which requires a study of China's energy needs is signed into law. This would delay a CFIUS review of CNOOC's bid.
10 August 2005	Unocal accepts Chevron's offer.

Source: Adapted from Nanto, Dick K., James K. Jackson, Wayne M. Morrison, and Lawrence Kumins. 2005. "China and the CNOOC bid for Unocal: Issues for Congress". *Congressional Research Service*, Washington, DC: The Library of Congress.

consistent with [free-market] principles".[36] American opponents argued that the only reason that CNOOC could make the bid was that

> $7 billion of the overall $20 billion costs (the $18.5 billion bid plus the assumption of some debt and a $500m 'kill' fee to Chevron if the latter is seen off) is coming via a parent entity from its ultimate owner – the government.[37]

Added to this was the fact that another $6 billion came from one of the state owned banks, and the total $11 billion was made available to CNOOC with little or no interest returns expected from it. The vice president of Chevron, Peter Robertson, perceived CNOOC's offer as unfair and argued that "We're not competing with a company ... we're competing with a government".[38] To many in the US, this seemed entirely outside the proper rules of free-market transactions as US companies would have had to pay much higher interest rates meaning that CNOOC was essentially getting money cheaper than its American counterparts.

However, counterarguments could suggest that US governmental subsidies negate the above criticisms of unfair business practices by the Chinese. While evidence suggests that the PRC was willing to offer $11 billion in financial aid, most of which was given as a loan and as such was expected to be paid back, "The [US] federal government spent $92 billion in direct and indirect subsidies to business and private-sector corporate entities – expenditures commonly referred to as 'corporate welfare' – in the fiscal year 2006".[39] This suggests that perceptions of unfair business practices of CNOOC and the Chinese are informed by particular anti-China agendas, as criticisms of unfair Chinese governmental help cannot stand in light of US federal aid to American corporations, the most notable being the $421.6 billion financial-crisis rescue which rescued two of America's three giant automakers from "the brink of collapse" in 2008.[40]

The bid was subject to significant debate in Congress and Senator Byron Dorgan (D-ND) raised concerns of reciprocity as well as national security when he stated that,

> This is a fairly significant issue and the fact is, we should deal with this in a manner that reflects our national interest.... Oil and gas are important strategic assets. We ought to delay this to examine the larger question. Do you think Unocal could buy Cnooc? Not in a million years. The Chinese government would not allow that.[41]

The fact that the Chinese company's offer was also made in cash, "$3 billion of it from its own balance sheet" only made the deal look more inequitable in the eyes of those who saw governmental backing of a corporate purchase to be against free-market principles.[42] I argue, however, that the perceived inequity of CNOOC's offer was based on how it was discursively represented. Just as American opposition to PRC financial backing of the bid seemed somewhat misplaced in light of the historical largesse of the US government with regard to American corporate interests, US opposition to CNOOC's bid on the grounds that PRC assistance gave it an unfair advantage due to the size of the Chinese government's coffers seems incongruous with the American business model which produced Wal-Mart, a "corporation [which] is too big to challenge".[43]

CNOOC was also viewed with suspicion by its critics as the scale of the bid raised questions as to China's real interest in the US oil company. Grant Clelland wrote that,

> Overshadowing all the political bluster surrounding the $18.5bn (£10.2bn, E15.2bn) bid by China National Offshore Oil Corporation (CNOOC) to buy America's Unocal is the question of whether CNOOC would be vastly overpaying for the oil and gas company ... if CNOOC should win this first-ever Chinese takeover battle for an American company, it would have to pay Chevron a $500m break-up fee and assume $1.6bn in debt, so the final price would be $20.6bn. That is nearly CNOOC's own market capitalization of $22bn.[44]

Feelings surfaced that surely, then, because Unocal could not be worth the $18.5 at face value, China must have had ulterior motives. Voicing such concerns, Rep. Thomas M. Reynolds (R-NY) stated that

> One has to wonder, why is CNOOC's bid so much higher? China is clearly offering a great deal more than Chevron's accepted bid.... Overpaying by this much – $2 billion – is illogical; unless the bidder has ulterior motives, such as gaining a measure of control over American energy supply.[45]

Concerns surrounding the strategic location of Unocal's assets and the would-be impact on US SOS if it were sold to CNOOC also stemmed from the CTD. Because much of Unocal's production was located in the South China Sea, opponents to the sale argued US national interests would be hurt as the United States would lose a vital foothold in an increasingly vital oil-producing region while China would be gifted an extremely strategic reserve which essentially lay on its doorstep as "Most of [Unocal's] holdings are in Asia, near China".[46] However, the discursive element of such readings again becomes evident as analyses of global oil production suggest that the market is fungible and generally tends to balance itself, negating the importance of its location. To this end, Hufbaur states that,

> If the oil market is indeed fungible, then Unocal production hypothetically directed to China under CNOOC's management would simply replace other imports that would have gone to China otherwise. Since overall global supply would remain the same, the price of oil would not be affected.[47]

Such arguments prove that alternative, evidence-led theories do exist which challenge the notion that Chinese acquisition of Unocal would harm US interests, and thus illustrate how the CTD affects readings of the bid.

Critics of CNOOC also believed that Unocal's expertise in the practice of deep-water drilling could be a primary driver for Chinese interest. Some American commentators suggested that Unocal's technology was dual-use and its availability to the Chinese would therefore counteract US military and political supremacy. In order to allay such fears, Fu Chengyu, the chairman of CNOOC, rebuffed these claims stating that "I have no talent for politics, only business".[48] It can once again be illustrated that the CTD and ESD worked to position China as a threat in this regard as had China not been placed in an antagonistic role to the United States, CNOOC's desire for such drilling technology would not have raised any ire as it would be within the logical remit of any growing upstream oil company.

For these reasons CNOOC's bid became highly politicized and debates surrounding the resultant impact on US interests raged. CNOOC's bid eventually went before Congress where, on 30 June, the House, led by Rep. Pombo, passed a resolution (H.Res. 344) calling for a CFIUS (Committee on Foreign Investment in the United States) review of CNOOC's proposal.[49] CFIUS, in conjunction with the attached Exon-Florio Amendment, "enables the US government to restrict, reject or impose conditions on foreign investments into the country on national security grounds", and this is especially relevant when the FDI involves a US economic sector which is part of the country's 'critical infrastructure'.[50] Rep. Carolyn Kilpatrick (D-MI) also led the House to include the Amendment H.Amdt. 431 to the appropriations bill (H.R. 3058) which prohibited the use of Treasury funds to approve CNOOC's purchase of Unocal, and this added yet another barrier to the Chinese company's bid.[51]

Chevron responded to CNOOC's offer, which had been submitted to CFIUS for review on 2 July, by increasing its bid to $17.1 billion on 20 July, realizing that if it did not make its own offer competitive with CNOOC's it would lose the deal.[52] Despite the increase, it remained $1.4 billion short of CNOOC's offer. In addition, Chevron's offer was not nearly as liquid as CNOOCs which was "all-cash upfront" but consisted instead of 40 percent cash and 60 percent shares.[53] Other attractive aspects of CNOOC's offer that Chevron did not match included CNOOC's

> commitment to retain the jobs of substantially all of Unocal's employees, opposed to Chevron's plan to lay off employees, especially in the United States. CNOOC also stated that it was willing to continue Unocal's practice of selling and marketing all or substantially all of the oil and gas produced from Unocal's U.S. properties in U.S. markets.[54]

However, despite the attractiveness of CNOOC's offer, and despite its protestations that its motives were entirely financial, the political pressure which was placed on the company led it to withdraw its bid on 2 August, six days before H.R. 6/P.L. 109–58 was signed into law. With CNOOC removed as a potential buyer, the board of Unocal approved Chevron's revised offer on 10 August.

4.4 Official discourse, CNOOC, and US national security

This section will examine the US government's position towards China in 2005 in order to position the CNOOC/Unocal debate within it. By examining the official discourse relating to the CNOOC/Unocal affair it will become clear that China had been systematically placed as a major potential challenger by the American political establishment.

One of the primary reasons CNOOC's bid for Unocal is so illuminating is because it highlights the inexorable links between energy acquisition, the economy, and security in modern international politics. Addressing the House on 30 June 2005, Rep. Bob Ney (R-OH) articulated these links when he stated that,

> at a time of rising prices on global oil supplies, ready access to energy resources is a vital element to our economic security. It is imperative that the United States protect its access to Unocal's energy resources in order to protect our economy and our national security.[55]

The unique uproar caused in Washington by CNOOC's bid led Kenneth Lieberthal to ask

> Do we see each other inevitably as antagonists, or do we see a world of globalization from which both sides benefit? This is the big issue.... And that framework, one way or another, will drive an enormous number of policy decisions.[56]

The answer lies in the way in which oil has become intrinsic to national security as well as in the way in which China has been cast as an antagonist to the United States, both by the political establishment, with those in government and opposition, as well as in the wider American context. Addressing the issue of US ES, Robert Samuelson writes that "We must defend our interests, but if we reflexively treat the Chinese as a threat, we will answer our own question: They will become a threat".[57] CNOOC's bid was not read as a simple commercial proposal as "most opposition to [CNOOC's bid] appears founded on emotive arguments about the supposed threat from China and Chinese companies, not the offer's commercial merits".[58] Although the CNOOC/Unocal affair represents an apogee of the way in which China was positioned as a threat to US interests, a process of institutionalization had seen this treatment become entrenched in the US establishment in the years prior to 2005. Because it was hotly debated in Congress, Congressional records surrounding CNOOC's Unocal bid are very illustrative. Although much of the discussion in the House reflected the unease with which America viewed a rapidly rising China[E] it must be remembered that these statements never reflected American perceptions of China in their entirety. A percentage of Americans supported the deal. Despite this, however, the deal fell

---

E   Congress voted 398–15 to back resolution HR344, sponsored by Rep. Richard Pombo, to veto the sale.

through as the CTD carried the day. The first step to understanding how the anti-Chinese atmosphere became so prevalent will be to examine the official US security strategy.

4.4.1 The United States' National Security Strategy of 2002, the Bush administration, and China

The *NSS* of 2002 outlined perceived threats to US national security as well as some broad strategies which could be implemented to mitigate them. Although much rhetoric within the *NSS* surrounded issues of terrorism and non-state extremist elements, primary focus remained on states, and although excluded from the 'axis of evil', the *NSS* placed significant focus on China. Although it was not maligned in the way Iran, Iraq, or North Korea were, China remained a potent symbol of, if not anti-, then at the very least un-Americanism, and in a climate where un-Americanism and anti-Americanism became increasingly interchangeable, perceptions of China regarded it, as they had done since Tiananmen, as a potential challenger to the US.

When he delivered the *NSS* in 2002, President Bush echoed certain sentiments of the first Bush administration when he expounded on what he believed American values and interests to be and argued for support of their adoption throughout volatile and 'non-Western' regions. Through his staunch support of a conservative tenor of American values, President George W. Bush contrasted America to those states not seen to share them. He declared that,

> The great struggles of the twentieth century between liberty and totalitarianism ended with a decisive victory for the forces of freedom – and a single sustainable model for national success: freedom, democracy, and free enterprise. In the twenty-first century, only nations that share a commitment to protecting basic human rights and guaranteeing political and economic freedom will be able to unleash the potential of their people and assure their future prosperity.[59]

In this statement, American ideals such as freedom and liberty were posited in such a way as to vilify those countries which did not prioritize them. In stating "Either you are with us, or you are with the terrorists",[60] President Bush created strict division where powers who did not directly contribute to strengthening US interests were clearly posited as enemies. This us/terrorist dichotomy served not only to alienate America's fiercest enemies, but also served to further distance the United States from nations it was already dangerously close to being estranged from. Although terrorism became a defining motif in post-9/11 US security discourse, it is interesting that nation-states remained so prominent in the *NSS*. The President stated that there was only one model which *countries* could use to thrive, and thus implicitly suggested that states remained the primary actors on the world stage. The *NSS* clearly juxtaposed the US to other countries, and one of the countries which it was most apt to damn was China. As Robert Kaplan writes,

Its military is an avid student of the competition [with the US], and a fast learner. It has growing increments of 'soft' power that demonstrate a particular gift for adaptation. While stateless terrorists fill security vacuums, the Chinese fill economic ones.[61]

Indeed, the CTD was so dominant in the Bush administration that prominent members made unveiled comments concerning the China threat. Gary Hart, former US senator and presidential candidate, who chaired the US Commission on National Security in the 21st Century, stated that Lynne Cheney, the vice president's wife, proclaimed that the greatest threat to US national security was China.[62] According to Hart, Cheney stated that "sooner or later the US would end up in a military showdown with the Chinese Communists. There was no avoiding it, and we would only make ourselves weaker by waiting".[63] Although the *NSS* served to identify the threats to American national security at large, the Bush administration placed America in direct opposition to China. To this end Hart stated that "I am convinced that if it had not been for 9/11, we would be in a military showdown with China today".[64]

While the CTD portrayed it as an increasingly powerful and effectual power, China was not perceived to be free or democratic. Its entry into the global marketplace in the 1980s had made it wealthy and it had gained power through that wealth, but few would argue that China was a supporter of wholesale free enterprise. Alan Tonelson, a research fellow at the US Business and Industry Council, elegantly summarized America's disquiet with China when he stated that "It is a system the likes of which we have never seen before because it consists of a combination of communist and free-market practices".[65] Digging further into the confusion over China, James McGregor writes that,

> our government's viewpoint on China is unfocused, fractured and often uninformed. Is China still the Red Menace of the Cold War or a hot new competitor out to eat our economic lunch? Both views as well as a hodgepodge of other interpretations can be found in the halls of the White House, Congress and the Pentagon. Add to that confusion a vicious domestic political culture that brooks no compromise, and the chances of formulating a coherent China policy approach nil.[66]

Such incongruences between China and the US coincided with other, more direct attacks on American notions of liberty and freedom. China remained widely condemned for its human rights abuses and its limitation of political freedoms; the conditions President Bush stated were necessary to "unleash the potential of their people and assure their future prosperity".[67] Perhaps to minimize the sting of these critiques the US President drew attention to what he considered to be steps taken by China towards becoming a responsible stakeholder in the international community. In his *NSS* address he stated that "Chinese leaders are discovering that economic freedom is the only source of national wealth. In time, they will find that social and political freedom is the only source of national greatness".[68] Despite his concession to the progress China had made, President Bush's compliments were barbed and served to continue the Othering process of China which the CTD stimulated. Indeed, the *NSS* states that,

> The United States' relationship with China is an important part of our strategy to promote a stable, peaceful, and prosperous Asia-Pacific region. We welcome the emergence of a strong, peaceful, and prosperous China. The democratic development of China is crucial to that future. Yet, a quarter century after beginning the process of shedding the worst features of the Communist legacy, China's leaders have not yet made the next series of fundamental choices about the character of their state. In pursuing advanced military capabilities that can threaten its neighbors in the Asia-Pacific region, China is following an outdated path that, in the end, will hamper its own pursuit of national greatness. In time, China will find that social and political freedom is the only source of that greatness.[69]

So no matter what progress had been made by China in its efforts to modernize and develop, American discourse, aligned with the Bush administration, positioned it as a country governed by an untrustworthy and authoritarian regime. In a sentiment echoed by harsher American critics, R. James Woolsey, former Director of Central Intelligence, described China as a "Communist dictatorship".[70]

In addressing the perception of US-Sino relations in the wider American political establishment, Francis Schortgen states that,

> the notion of competition and challenge [with China] has never fully dissipated. Indeed, over the course of the 1990s and into the twenty-first century, a persistent ebb and flow of fixation on the China challenge took center stage. Defined in both military terms and trade considerations, China was rarely out of the spotlight.[71]

Robert Samuelson notes that

> We cannot decide whether China is a threat or an opportunity, and until we do every discussion of our relations seems to slide into confusion and acrimony. The latest example is the noisy controversy over the bid by CNOOC Ltd. to buy the American oil company Unocal Corp.[72]

Clearly then, despite China's efforts to rehabilitate its image in the wake of its Tiananmen crackdown in 1989, the CTD held sway on US perceptions of China so that China remained vilified to a degree that any attempted Unocal takeover by CNOOC was seen to be inappropriate to American sensibilities. To this end, Katherine Griffiths writes that, "Summing up some of the hysteria that CNOOC's offer has caused among some on Capitol Hill and New York's financial community, one television commentator yesterday said the bid was 'blood money on the backs of students from Tiananmen Square'".[73] The CTD promoted acrimonious readings of China and thus helped to shape US national security policy towards it.

The NSS: projection of impending Sino-American relations

Although issues relating to ES are pertinent to aspects of the *NSS*, anti-China sentiments are primarily viewed through the CTD rather than the ESD. Towards the close of his address on the *NSS* President Bush commented directly on the business links between China and the United States. He managed again to admonish China while simultaneously offering hope concerning the future of Sino-American economic relations. With bilateral trade amounting to $100 billion, and with China being America's fourth-largest trading partner in 2002, President Bush stressed China's importance as a trade partner.[74] Others, including James Dorn of the CATO Institute, argued that its ascension to the WTO in December 2001 would make China have "a willingness to normalize trade relations" and help make it a more responsible stakeholder.[75] The fact that US elites even sought to normalize China is proof that China was perceived as *ab*normal. Despite American willingness to engage China, the President refrained from full praise of Chinese progress and undercut much of the optimism he had previously outlined when he stated that,

> There are, however, other areas in which we have profound disagreements. Our commitment to the self-defense of Taiwan under the Taiwan Relations Act is one. Human rights is another. We expect China to adhere to its non-proliferation commitments. We will work to narrow differences where they exist, but not allow them to preclude cooperation where we agree.[76]

CNOOC's attempted purchase of Unocal certainly tested this spirit of cooperation three years later.

Although not the primary focus of the *NSS* of 2002, China had officially been singled out by the American government to be a rising power with interests that ran counter to those of the United States. Certainly, in the post-Cold War, and ensuing post-9/11 era, extremism and terrorism have been highly visible targets of American defence strategies, but Robert Kaplan argued that this was a distraction and that, "The Middle East is just a blip. The American military contest with China in the Pacific will define the twenty-first century. And China will be a more formidable adversary than Russia ever was".[77]

The importance of the *NSS* is found in the manner in which it both reflected and reinforced anti-Chinese sentiments and reinforced the CTD which impacted on the events of CNOOC's proposed takeover of Unocal three years later. The effect of the *NSS* is threefold. First, despite popular perceptions that America's greatest threat in the post-9/11 era emanated from non-state terrorist actors, the *NSS* actually accorded primacy to relations between states as a matter of US national security. Second, although not vilified in the extreme as a member of the "axis of evil", China had been deliberately singled out in the strategy as one of the states with which the US had to exercise caution in its dealings as China had yet to embrace fundamental aspects of a 'free' society as envisaged by American standards. Finally, due to its unprecedented growth, ES was tipped to be a main point of contention in the US-Sino relationship as China continued to develop. Therefore, although the *NSS* painted China as an antagonist with a wide brush, it positioned it so that subsequent interactions with China would be influenced by the CTD from the outset. An examination of Congressional debates during the 109th Congress in 2005 will support the argument that China was purposefully positioned as a threat by the American political establishment.

4.4.2 *The CNOOC/Unocal affair and Congress*

In 2005 the CNOOC/Unocal affair became a prime topic of debate in Congress with Senate and House sessions both devoted to the matter. The fact that there was widespread concurrence within Congress about how to deal with CNOOC's bid is quite remarkable as it is well documented that not only is there frequent antipathy between the executive and legislative branches of American government, but friction often defines the partisan relations within the legislative branch due to "a long-term polarization trend in the House that began in the early 1980s".[78] Because of partisan loyalties, and because approval is required from both the House and Senate in order to pass any legislation, Congress is often seen to be a cumbersome instrument of American politics. However, as a testament to the degree to which both central discourses had penetrated American politics, despite the partisan nature of Congress, dissent was muted with regard to the CNOOC/Unocal case as the House and Senate were conspicuously in agreement with one another that the takeover would pose a threat to US national security.

Congress and the CFIUS review

Congress directed CFIUS to review the bid in order to create a legally supported political barrier to CNOOC. CFIUS, created in 1975 by the Ford administration, was given the power to enact a review process of foreign acquisitions of US companies by Ronald Reagan in 1988 through the Exon-Florio Amendment to the Defense Production Act of 1950. This amendment increased the powers of the committee and was a direct result of the Japanese Yellow Peril of the 1980s.[79] The amendment was attached in the wake of Fujitsu's attempted merger of Fairchild Industries when "The debate over the Fujitsu merger fuelled the larger controversy over rising Japanese investment in the United States".[80] It is remarkable how similar this anti-Japanese sentiment was to the later anti-China sentiment in positioning Asia as an Other to the United States. While Japanese corporate expansion defined the Yellow Peril of the 1980s, Chinese corporate expansion defines it in the 2000s and "If there's an asset up for sale anywhere in the world, people are looking to China [to buy it]".[81] With the CTD firmly entrenched, Chinese firms have now become a primary target of CFIUS attention.

CFIUS is run by representatives of 12 government agencies including the Department of Energy, Homeland Security, and Defense. The inclusion of these three departments gives some indication as to the role and nature of CFIUS. The review process begins with a 30-day investigative authorization into the proposed transaction, after which the committee is granted another 45 days in which to permit the acquisition, but the President ultimately has the opportunity to block the proposed transaction. Oftentimes, foreign companies withdraw their bids before they begin the review process either because they genuinely believe that the American political establishment will see the bid as a contest to US interests, or because the whole process is so overwhelming that it is financially necessary to withdraw. By 2005, of more than 1500 filings, only about a dozen had gone through both review processes and reached the President for a final decision. Only once had a deal been blocked, when Mamco Manufacturing, a US aero parts maker, was bid on by the China National Aero-Technology Import and Export Corporation in 1990.[82] The fact that the only deal ever directly blocked through CFIUS involved a Chinese company surely set a precedent as to future bids and set the tone of the Congressional environment in which the CNOOC bid for Unocal was received. It has been noted that "The US security agencies within CFIUS are less likely to offer flexibility when it comes to Chinese investment. Whether or not it is justified, the US government considers Chinese investment to present special concerns".[83]

The invocation of a CFIUS review was designed to create a situation in which CNOOC was directly prevented from purchasing Unocal through imposed legislation, or would create an environment in which so many procedures and assessments would be attached to CNOOC's proposal that it would have to be withdrawn due to matters of practicality. The disruptiveness of CFIUS reviews can be so great that oftentimes "bidders drop their offers when confronted with the committee review and its conditions" rather than attempt to challenge the committee and justify their commercial intent.[84] Essentially, this strategy aimed to drown CNOOC in paperwork and unsettle the shareholders of Unocal who would otherwise be pleased to accept the higher offer which CNOOC tabled over their Chevron counterparts. Congressional action had the desired effect, much to the chagrin of CNOOC and the outright anger of the Chinese government. Explaining its withdrawal from its bid for Unocal, CNOOC released a rather muted statement explaining that, "CNOOC has given active consideration to further improving the terms of its offer, and would have done so but for the political environment in the US".[85] The anger of Chinese authorities, however, contrasted sharply with the quite subdued nature of CNOOC's official statement. The Chinese foreign ministry released a vitriolic statement stating that,

We demand that the US Congress correct its mistaken ways of politicizing economic and trade issues and stop interfering in the normal commercial exchanges between enterprises of the two countries ... CNOOC's bid to take over Unocal is a normal commercial activity between enterprises and should not fall victim to political interference. The development of economic and trade co-operation between China and the United States conforms to the interests of both sides.[86]

In the end, however, the threat of CFIUS review by the US Congress was successful as it was able to create specific barriers to CNOOC's transaction, as well as help to exploit a general climate of Sinophobia in America which had been promoted through the CTD. Tim Payne, a spokesman for CNOOC, explained that the hostile environment created "a level of uncertainty that presents an unacceptable risk to our ability to secure this transaction", and he went on to vent some frustration when he added, "Are we pissed off? Yes".[87]

Outlining House Resolution 344–109th Congress, 1st Session

Whereas anti-China sentiment, through the CTD, had been vocal but rather disorganized in the years prior to 2005, the ESD helped to focus particular readings of Chinese ES through reference to the CNOOC/Unocal affair. The most visible political result of this was the passing of H.Res. 344.[88] In order to exhibit the degree to which CNOOC's bid garnered the attention of US Representatives, H.Res. 344 should be referenced in its entirety as there is no way to economize on the implications it lays out regarding the ESD and its direct impact on anti-Chinese sentiments.[F]

This bill exhaustively addressed those reasons behind what some saw as Congress' decision to interfere with the free market and sabotage CNOOC's bid, but what most saw as justifiable measures designed to protect America's national security. These reasons were significantly and visibly influenced by the CTD and the ESD. Those who were against the sale, which included a significant majority in Congress, argued that the bid should be blocked because the national security of the United States would be jeopardized if strategic American assets, in this case an oil company, were sold to foreign competitors; especially China. Essentially, "The claims [that CNOOC's purchase of Unocal would hurt US interests] stemmed from three central facts: CNOOC is a foreign company; the Chinese government controls it; and it has unfair financial support from the Chinese government".[89] Thus, opposition to CNOOC's purchase was not only based on the fact that it was foreign, but specific antipathy was based on the fact that it was Chinese. However, the 'foreign' element of the argument is highly questionable in light of the many foreign, yet admittedly Western, corporate acquisitions of US oil interests in the years directly preceding CNOOC's bid for Unocal, and which were outlined in section 4.2. In light of this I argue that reference to 'foreign' investment simply served to legitimize specific arguments against Chinese investment.

American wariness of China's ability to threaten US interests clashed with the US need to engage with China with which it had entered into mutual most-favoured-nation (MFN) treatment in 1980, and which remained active despite

F   See Appendix 1.

controversies in the 1980s and 1990s.[90] This conflict, specifically as it relates to CNOOC's bid for Unocal which, according to H.Res. 344, "threaten[ed] to impair the national security of the United States",[91] was clearly elucidated by Rep. Kilpatrick when she stated on the floor that, "China is an economic and military power. They are one of our largest competitors.... Should we work with China? Yes, we should. Should we turn over our government business to China? No, we should not".[92] The description of national security as it was used in Congress was intentionally left vague by many who invoked it so that it could encompass many factors which would not limit the scope of US authorities' jurisdiction when dealing with perceived threats.[93] Through various bills and resolutions, as well as the initiation of a CFIUS review, Congress effectively exploited notions of national security to vilify CNOOC's Unocal bid.

Foreign policy is based on ideas of 'national interests' and 'national security' which may have different meanings for different people, or even no precise meaning at all, "Thus, while appearing to offer guidance and a basis for broad consensus they may be permitting everyone to label whatever policy he favors with an attractive and possibly deceptive name".[94] It is essential, therefore, that the elements of these terms are clearly defined because 'national security' and 'national interests' are similar yet not identical. National interests can supersede national security in that the latter is a constitutive part of the former, and interests ultimately encompass more elements than security alone, although security is often considered to be *the* essential element. This long-standing preoccupation with security, particularly negative security, can be evidenced by Wolfers who stated, "Statesmen, publicists and scholars who wish to be considered realists, as many do today, are inclined to insist that the foreign policy they advocate is dictated by the national interest, more specifically by the national security interest".[95]

Nanto suggests that national interests can be divided into three component parts, security (i.e. the protection of property and life), prosperity (i.e. the protection of economic welfare and commerce), and values.[G,96] The sale of Unocal to CNOOC may be examined in this context as H.Res. 344 opposed Chinese ownership of Unocal by employing these three component parts of national interests. The arguments made are informed by the CTD in conjunction with the ESD and suggest that US security, prosperity, and values would have been challenged if Unocal were sold to CNOOC.

4.5 CNOOC, and the protection of American 'national interests'

Having explored how CNOOC's bid was received and discussed in official US discourse it is useful to now examine how the central discourses were employed

---

G President Bush summarizes his conceptions of American values in his outline of the *National Security Strategy* of 2002. These include notions such as liberty, democratic freedoms, and human rights.

in wider non-official discourse in relation to the CNOOC/Unocal affair. Although reference will be made to the American political establishment, including Congressional records, this section looks towards wider, non-elite discourse as well. By establishing the discussion of the case study on the three themes with relate to national interests (security, prosperity, and the preservation of values) it will be demonstrated how the central discourses played a discursive role in CNOOC's failed bid for Unocal, and also how the ESD and CTD functioned together to promote a specific reading of China.

4.5.1 Security: the protection of property and life

China threat arguments rest predominantly upon the pillar of security, which is essential to US prosperity and the preservations of its values. Issues of resource scarcity, Chinese expansionism, and militarization, including the access to dual-use technologies, will be examined in order to understand perceptions of China as a security threat to the US. In their letter urging President Bush to block CNOOC's bid, Representatives Pombo and Hunter wrote that, "As the world energy landscape shifts, we believe that it is critical to understand the implications for American interests and most especially, the threat posed by China's governmental pursuit of world energy resources".[97] The scarcity of energy resources which would be sold to China, as well as provision of dual-use technology to China, a country in the midst of substantial military modernization, provided grounds for criticism of CNOOC's bid. Both invoke visions of Kent Calder's 'deadly triangle' between economic growth, energy shortage, and militarization in East Asia.[98]

Security and scarcity: SOS

H.Res. 344 linked US security to oil scarcity by stating that the global demand for oil was at its highest point in history in 2005 while global production was also the lowest in history. The high demand combined with the low production of 2005 created an unstable environment in which oil became increasingly more valuable as states competed for SOS. William A. Reinsch, president of the National Foreign Trade Council and former trade official in the Clinton administration stated that "The national security argument is a fair one.... When you talk about energy supplies, and the market is tight, there is a national security issue. You are going to have a lot of people pounding the table".[99]

Regarding the sale of domestic resources to the Chinese,[H] Larry M. Wortzel, a former military attaché to the American Embassy in Beijing and a member of the Congressional United States-China Economic and Security Review Commission said that, "A lot of this is corporate self-interest and local politics, but there is also a legitimate question to ask.... Do we want a foreign power, whose military intentions in the long term are not clear, to own energy assets inside our

H   Unocal had production in the US.

border?"[100] Although the CNOOC/Unocal affair helped to focus anti-Chinese sentiments in the US, the issue extends beyond CNOOC and Unocal. Jaffe writes,

> there is a real reason for American concern about China's suddenly voracious oil thirst. Right now, that thirst translates into a willingness to overbid for assets like Unocal. But to what strategies might China turn if Western competitors prevent it from acquiring choice assets?[101]

The implication raised by Jaffe is that China may turn to a more aggressive and militaristic stance in order to secure its energy requirements if the market will not allow it to do so peacefully.

While some argued that the CNOOC bid was an economic issue with economic answers, to many others, ES was too important an issue to leave to market mechanisms. Michael R. Wessel, a member of the Congressional commission said, "I think most people would agree that oil is a national security issue. What is still to be determined, of course, is what to do about it".[102] To those who saw CNOOC's Unocal bid as an affront to US national security and SOS, the solution was clear; block the sale. Although the oil market is often argued to be the most fungible global market, concerns were raised by those such as members of the House Armed Services Committee that China was attempting to bypass the market through mercantilist strategies.[103]

Because China was somewhat of a late entrant into the geostrategic oil game, it was perceived by some that China was eager to establish itself wherever possible as certain regions were already dominated by the West. In this scenario, Chinese gains in the zero-sum SOS could destabilize the United States. To this end Alf Young writes that

> [China] longs for greater security of supply every bit as much as the largest importer of all, the United States. But it knows that access to some of the riches oil provinces, like the Middle East, is already controlled by the West.[104]

Arguments emerged that by denying China access to legitimate sources of energy, the US forced China to turn towards producers such as Sudan and Iran, which raised considerable security concerns as well and will be addressed below. Moreover, China was eager to acquire Unocal's resources as China's own production was stagnant, and it was increasingly relying on oil imports as it had been a net oil importer since 1993.[105]

It was also argued that despite the fact Unocal is based and has significant production in the United States, its acquisition by China would be a major strategic coup as Unocal's assets extend beyond US borders and include many assets in Southeast Asia. Were CNOOC to be successful in its takeover of Unocal, China would have access to resources much closer to its own sphere of influence. This would also mean that China would also be increasingly able to rely on its own navy to protect its interests which would mean that China would rely less on the US Navy's protection of vital sea-lanes which would reduce US leverage over China.[106]

Importantly, the issue of scarcity is also inexorably linked with US perceptions of insecurity stemming from the modernization and development of China's military capabilities, especially its navy. Its ambitions are illustrated by the idea that

> China may want to eventually build a series of naval and other military bases in the Indian Ocean – a so-called 'string of pearls' – so as to support Chinese naval operations along the sea line of communications linking China to Persian Gulf oil sources.[107]

As oil and gas become increasingly scarce, China threat adherents argue that China will increasingly turn away from the market to provide for their energy needs, and will instead find military solutions to own oil and gas at the wellhead. Even if China's military modernization and growth were predominantly predicated on being able to secure its own energy imports, it would have serious knock-on effects with regards to issues such as Taiwan, its relationship with Japan, and global stability. Rep. Barton (R-TX), chairman of the House Energy and Commerce Committee, and Rep. Ralph Hall (D-TX), chairman of the subcommittee on energy and air quality, wrote a letter to President Bush on 27 June 2005 in which they stated that the ability for the US to secure its oil and gas needs is "threatened by China's aggressive tactics to lock up energy supplies around the world that are largely dedicated for their own use".[108]

Militarization and expansionism

The ESD has also been used to legitimate the China threat in relation to Chinese military growth. Rather than viewing Chinese military modernization as a normal step towards development, as was demonstrated in Chapter 2, US proponents of the CTD have traditionally viewed PLA modernization with extreme suspicion. Whereas PLA modernization was perceived to be directed towards Taiwan in the 1990s,[109] PLA modernization in the 2000s was perceived to be directed towards facilitating China's ES as "China's interests increasingly extend beyond its shores to resource-rich areas of the developing world and the trade- and energy-choked SLOCs".[110] Although many current debates surrounding China's military modernization and expansion are linked to Chinese SOS, this modernization and expansion can then itself be perceived as threatening when it is decoupled from ES issues. When ES is not invoked to qualify China's military modernization and expansion, it is perceived by those in the US to look outright aggressive as "This is their new policy of deterrence.... They want to show the U.S ... their muscle".[111] Because China's military growth is dependent on SOS of oil and gas, and because CTD proponents read Chinese action through ESD optics, situations such as the CNOOC/Unocal affair become hot-button issues.

Furthermore, Chinese policies aimed at redressing its insecurity of supply are perceived as upsetting the status quo in the Western Pacific and America's place within it. The ability of the United States Navy to project force in a truly global manner means that it is currently the only power which is able to patrol the SLOCs through which a majority of the world's petroleum flows. Importantly then, the US is not only able to protect its own supplies, but it is also theoretically able to disrupt those of any competitors. For this reason, China, driven by the scarcity notion of the ESD, has embarked on an enormous naval build-up and modernization in order to create a true 'blue water' fleet to protect maritime chokepoints, such as the Malacca Strait, through which 80 percent of China's imported oil flows.[112] For this reason, "China has sought to increase its military and diplomatic presence in the South China Sea and beyond. Following a 'string of pearls' policy, it has sought access to bases in Pakistan, Bangladesh, Burma and Cambodia" as well as Indonesia.[113] With the same spirit which it is instilling in its NOCs to secure oil from abroad, the Chinese 'go out strategy' is being applied to its navy. Thus, the role of perceptions is made clear as the ESD and CTD work to portray Chinese naval modernization and expansion as aggressive when it could otherwise be perceived to be consistent with normal processes of modernization.

China's claim for Taiwan has become entangled in arguments surrounding China's ES strategies as well. China has never been unambiguous as to its interest in Taiwan, and, problematically, the United States has been equally unambiguous as to its intent to protect Taiwan from any mainland aggression.[114] However, diplomatic and geopolitical considerations aside, China simply does not have the maritime capabilities to invade Taiwan, leading CTD proponents to be wary of the modernization which would enable China to patrol international sea lanes as this could allow China to overcome the logistical problem of transporting troops across the Taiwan Strait en masse. In addition, with specific reference to Unocal, Rep. Hunter, chairman of the House Armed Service Committee, "suggested that China could use its ownership [of Unocal] as leverage if it decided to invade Taiwan".[115] The idea that CNOOC ownership of a middling oil company such as Unocal could play a significant role in the US projection of power may be somewhat questionable, but the fact that it can be considered at all is significant in that it provides evidence as to how deep the CNOOC/Unocal affair and the central discourses have become embedded in American public discourse on China.

Although it is reasonable to expect any country to want to secure its own energy interests, China's naval development has not been widely embraced by others. Michael R. Wessel, a member of the US–China Economic and Security Review Commission, argues that the logic that may be applied to Western powers does not necessarily apply to China. He states that China is "not a market economy – that's the real challenge we have here.... They see resource acquisition as an integral part of their military plans".[116] Clearly then, the relationship between Chinese ES and its militarization becomes confused by US observers as they portray Chinese military enhancement as both a result of, and a reason for China's ES policies.

The transfer of dual-use technologies and the impact on US national security

The CNOOC/Unocal issue also raised questions surrounding the transfer of sensitive technologies to China. H.Res. 344 addresses at least three major concerns to this end. Richard Pombo, whose district included the headquarters of Unocal, was instrumental in passing the resolution which stated that Chinese ownership of Unocal would "threaten to impair the national security of the United States", and he believed that a prime indicator of that threat was the dual-use nature of the commercial technologies owned by Unocal.[117] Another vocal opponent to the sale of Unocal to CNOOC was Rep. Nancy Pelosi who raised her concerns about the Chinese acquisition of dual-use 'cavitation' technology that would have been part of the deal. She stated before Congress that

> Cavitation is a process which Unocal uses to go into deep water to drill for oil. That same technology can be used by the Chinese to do nuclear tests underground and to mask them so we would not ever be able to detect them.... Given China's commitment to improving its military capabilities, why would the United States permit the sale of this kind of technology? Left on its own, we probably would not. But as part of the UNOCAL deal, it is being pulled through with this Trojan horse.[118]

Pelosi makes the argument that because the US government would not provide China with the technology and expertise for military cavitation, the sale of Unocal to Beijing should be blocked on the grounds that the technology and expertise for civil cavitation could easily be adapted to military use. Despite their intended application to oil exploration, these sensitive technologies, such as seismic analysis and processing, downhole logging sensors, and modelling software, could enhance the PLA's threat to US forces without it having to go through a lengthy domestic modernization process. The argument follows that access to Unocal technologies would allow China to bypass a lengthy process of modernization which it would otherwise have to proceed through in a slow and organic manner. It is for this reason that such technologies require export licensing; licensing that the US would otherwise deny China. Regarding these restrictions, Clyde V. Prestowitz, president of the Economic Strategy Institute in Washington and a trade official in the Reagan administration, stated that "as a rising military power, China is not viewed as a strategic partner but as a strategic competitor".[119] With specific reference to CNOOC's bid for Unocal, Prestowitz also stated that "it does raise the issue of whether this gives influence or some kind of potential importance to a government that may not always be friendly to us".[120] As well, it is not only the technology and experience of Unocal which could be adapted for military use by the Chinese, but also the materiel. Carolyn Bartholomew explains that,

> Some oil exploration and drilling equipment, including software, is controlled for export because of its dual-use potential, in some case for nuclear

testing, in others for detecting submarines. Unocal is also the owner of the last U.S. source of rare earth minerals. Rare earth minerals are a critical component of magnets used in JDAMs, Smart Bomb technology, and other vitally important military applications.[121]

Thus, the argument follows that by selling Unocal to CNOOC, the United States would be divesting itself of direct access to these rare earth minerals and it would have to begin to rely on the global market, which in times of war could be disrupted. Moreover, China already has vast reserves of rare earth minerals[1] and would therefore be given another great strategic advantage over the United States.

China's involvement with regimes such as Iran and Sudan also served to raise the ire of the United States when CNOOC bid for Unocal. Those who disputed the logic of disallowing Chinese ownership of an American oil company argued that if doors to the American oil marked were closed to China it would only serve to strengthen ties between China and oil-rich rogue states.[122] Such ties would help legitimize, or at least sustain, these regimes whose existence is problematic for the US in its global ambitions. Indeed, in some cases China has purposefully undermined international sanctions which had been placed on these countries.[123] However, the louder argument was made by those who believed that the sale of Unocal to CNOOC would, in effect, grant access of sensitive military technologies, including nuclear secrets, to Khartoum and Tehran via Beijing.[124] Essentially, because Chinese companies are active in these rogue states, there is no guarantee that these dual-use technologies would not be sold to them; in this scenario the US simply could not enforce export controls or sanctions. Wayne Morrison states that "On June 28, 2005, the House passed an amendment (H. Amdt 381 to H.R. 3057) that would prohibit the U.S. Export–Import Bank from financing the sale of U.S. nuclear power equipment to China".[125] It was argued that a provision such as this could be easily circumvented by China's acquisition of Unocal. Therefore, with regard to the possibility of granting dual-use technologies to competitors, blocking the sale of Unocal to CNOOC was seen to be the lesser of two evils.

Despite protestations from the Chinese government stating that the PRC had no interest in acquiring Unocal for any technology that might lend itself to military applications, there were those in Washington who remained extremely wary of Chinese intentions as Unocal seemed to be worth far less as a simple economic asset than CNOOC was offering (see section 4.3). CTD proponents perceived Chinese interest in Unocal to be based on issues larger than its commercial value alone. Many argued that the money being offered for Unocal did give insight into grave national security issues as the bid simply did not make sense as a commercial venture alone. Oil is a strategic asset and Rep. Pombo said he wanted to "send a message to the administration about the depth of

---

1 The US Geological Survey estimates China produces 95 percent of the world's output of rare earth elements ("Facts of China's Rare Earth Reserves", 2011).

Congressional concern".[126] Because Chinese companies had been successful with other commercial acquisitions, blocking the sale of Unocal to CNOOC did not seem to conform to the rules of fair play. Pombo, however,

> said he saw no public policy issue with the recent $2.5 billion bid by Haier, a Chinese appliance maker, for Maytag, which he regarded as similar to the Japanese buying Rockefeller Center at the end of the 1980's ... I didn't like it, but it was no big deal.[127]

Selling an oil company, however, was perceived to be dangerous.

4.5.2 Prosperity and reciprocity: the protection of economic welfare and commerce

The prosperity of the Nation was another factor which was appealed to by critics who argued that CNOOC's purchase of Unocal would harm US national interests. Richard Pombo and Duncan Hunter wrote to President Bush arguing that

> such an acquisition raises many concerns about U.S. jobs, energy production and energy security.... We fear that American companies will find it increasingly difficult to compete against China's state-owned and/or controlled energy companies, given their mandates to supply China's ever-growing demand for energy, which will increasingly need to come from foreign sources.[128]

H.Res. 344 addressed three concerns surrounding the threat to US prosperity; a lack of reciprocity for US investment in China, unfair business practices on the part of CNOOC and the PRC, and the lack of an open market in China in which to sell American goods. Although these three concerns have been explicitly highlighted by the resolution, they point to a fourth, less-explicitly articulated concern as to China's relative economic growth to that of the US.

Lack of reciprocity and unfair business practices

With regard to reciprocity,

> Much of the congressional opposition to the attempted acquisition has swirled around questions of whether the links between CNOOC and the Chinese government provide it unfair advantages and whether the company is acting as a commercial competitor or an arm of the Chinese state in pursuing the deal.[129]

To this end, Rep. Barton stated that "if Unocal was trying to buy the Chinese National Offshore Oil Company, they could not do it, because Chinese law does not allow a foreign company to have controlling interest in a company in

China".[130] Charles Schumer detailed the various barriers that existed to US investments in Chinese companies and explained that PRC regulations state that foreign acquisitions of Chinese companies can only take place through equity or asset acquisitions which have to be approved by a board meeting of the Chinese company as well as through a shareholder's meeting.[131] As the Chinese government is the largest shareholder in CNOOC, as well as CNPC, and Sinopec, American critics of CNOOC's bid claim that it could flatly refuse any foreign approach as "any takeover of a major Chinese oil company would require approval from Beijing".[132] Chevron would therefore, hypothetically, be prevented from any purchase of a Chinese company such as CNPC, Sinopec, or CNOOC. Schumer was so incensed by the Chinese bid that he sponsored a bill which would impose a tariff of 27.5 percent on China's imports to redress the inequity of its currency policy.[133] He went on to rhetorically ask, "does anybody honestly believe that the Chinese would ever let an American company take over a Chinese company?"[134] Although it was certainly not a purchase which had a major impact on national security, American-based Anheuser-Busch's success in buying China's Harbin and Tsingtao breweries does suggest that the Chinese would not be as closed to American investment as some might contend,[135] and this helps to highlight the anti-China agenda of those who aimed to block CNOOC's purchase.

It was argued that Chevron, and thus the US, was also put at an unfair disadvantage due to the government backing of the Chinese company. Essentially, it was said, "Congressional concern is driven in part by the perception that China does not play by the rules in international trade policy".[136] Through government subsidies and low- to no-interest loans given to Unocal, American opponents said that CNOOC had access to 'cheaper' money than its US counterpart and it was, therefore, circumventing the rules of a free market economy. Peter Robertson, the vice chairman of Chevron in 2005, expressed his frustration at what he perceived to be the unfair business practices of CNOOC when he stated that, "Clearly this is not a commercial transaction.... We are competing with the Chinese government, and I think that is wrong".[137] Robertson also appealed to US sensibilities regarding free market transactions when he said that the CNOOC bid was "not fair trading", and added that Chevron would actually produce more oil and gas making their bid fundamentally more attractive.[138] Rep. Pelosi called CNOOC's bid for Unocal "a graphic example of America's energy vulnerability" and stated that "The Chinese government's control of CNOOC made the bid possible, not the free market".[139] Commenting on the perception that CNOOC was engaging in unfair business practices, Rep. Jefferson stated that because of government backing, "The fact of this is this is not a free market transaction ... and it puts every other competitor for the assets that they are seeking to acquire at a disadvantage".[140] However, arguments that CCP aid to CNOOC was unfair and by association 'un-American' can be deflated when one looks at the 2009 government bailout of the American auto sector as "Honda, Toyota, Nissan, Kia, Hyundai, BMW, and other foreign nameplate producers" who were not seeking handouts were "implicitly taxed when their weaker competition [was] subsidized".[141] Certainly, had laws of the free market prevailed, the American auto industry would have imploded during the 2008 financial crisis. European and Japanese automakers could have claimed that the American government was directly interfering in the free market in a similar fashion to what the US claimed the PRC was doing in 2005. Thus, stating that government backing of a commercial deal is unfair can certainly be put into question and also serves to highlight particular anti-China bias.

The situation could also be seen to be ironic when one considers that the Americans who criticized CNOOC's deal as unfair on the grounds that the Chinese government interfered in the bid also pleaded for American governmental interference. Claims that assistance by the Chinese government rendered CNOOC's bid fundamentally unfair were twinned with demands that the US government intervene in the transaction. Robin West, the chairman of PFC Energy, an oil consultant, explained the paradox when he stated that

> There are a lot of people in Washington who are really torn.... They believe in open markets and don't want to exacerbate matters with China. Yet, do you want a Chinese company that doesn't play by American rules to take advantage of American rules and get an American company?[142]

Regarding PRC subsidization of CNOOC, Sen. Evan Bayh (D-IN) stated that

> this renewed bid heightens my concerns about the heavily subsidized nature of CNOOC's financing. When foreign firms compete for assets in the U.S., it is essential that they do so on a level playing field with U.S. companies. Government subsidies tilt this playing field, and in doing so distort competition.[143]

Pombo added that "If the Chinese are willing to tell the Congress of a free nation to get lost, what assurance do we have that they wouldn't tell the free market to butt out too? I think the answer is 'none'".[144]

Finally, the negative impact on the American economy due to the inability to meet its SOS prompted other concerns about Unocal's sale to CNOOC. Because Unocal was relatively small its sale to CNOOC would not have significantly impacted on America's SOS in and of itself, although it would have had a massive impact on the precedent set. H.Res. 344 highlights the phenomenal growth in China's energy demand and stated that its consumption of crude accounted for more than one-third of global demand in 2004 and was set to grow even larger. H.Res. 344 suggested that this growth in demand would ultimately push the price of oil higher, and thus hurt the US economy. Others in Congress also argued that despite the promises made by CNOOC to continue to sell the oil produced by Unocal in the Gulf of Mexico to the US market, it made no such promises regarding its other global assets. Critics of the deal argued that China's entry into the upstream levels of the global oil market could reduce the fungibility of the market itself and ensure that the oil produced by CNOOC was directly shipped to China, bypassing the US market completely. Therefore, critics argued, the sale of Unocal would also have the added effect of weakening the United States' ability to ensure that the global oil market would conform to US law.[145]

Lack of open market in China

The lack of an open market in which to sell US goods in China transcends the CNOOC/Unocal debate and had been a long-standing point of contention in Sino-American relations. The current trade surplus between the US and China has resulted in an enormous trade imbalance in which US goods are effectively denied access to the Chinese market. Although many in the US felt that CNOOC's bid was the last straw in the inequitable economic relationship with China, some argued that blocking the bid could be a very dangerous move. Rep. Jim Moran (R-VA) stated that "They are holding a financial guillotine over the neck of our economy, and they will drop that if we do things like this that are not well considered".[146]

It interesting how a substantive argument by seventeenth-century China critics has been entirely upended by modern China's rise. As discussed in Chapter 2, Adam Smith was derisive of China as he saw it as idle and wasteful because it was bestowed with great natural wealth and did little to cultivate it. However, in the twenty-first century, any reference to China as 'idle' is absurd. Instead, it is China's commercial zeal which is now drawing complaints from the West as its economic power has surged. The vast natural wealth of China as described by Smith is also hardly a fitting description of contemporary China as its geography simply cannot supply it with the resources its growth demands. This, in turn, has led companies like CNOOC to look beyond Chinese borders for the supplies the country needs, and turned China's temporal progress into a threat.

China, Inc.: China as a growing economic competitor

It is possible to suggest particular economic challenges to US national security as they were outlined in H.Res. 344 while maintaining some political sensitivity and tact. It would, however, be wholly inappropriate in such a document to simply state that China's economic growth poses a threat to US interests. However, I argue that such a sentiment was strongly held and helped inform the debate surrounding CNOOC's bid for Unocal. Steve Lohr illustrates how perceptions of China have changed when he writes that now "China is both an engine of economic globalization and an emerging military power. In symbolic shorthand, it is Wal-Mart with an army".[147] The China which was feminized and dominated for much of its recent history has been replaced with one that is keen to flex its muscle, and it is this China which seems so alien to powers which have become accustomed to Chinese supplication. Nowhere can this change be seen more than in the general robustness of its economy.

Few countries can claim to have exhibited roughly 10 percent economic growth per year for a decade, but by 2005 this is precisely what China had done. During the same period US GDP growth only ranged from a high of 4.9 percent in 1999 to a low of 1.1 percent in 2001 which contrasts greatly with that of China, whose lowest GDP growth in those years was 7.6 percent in 1999.[148] US fears of China's rise are, therefore, more to do with growing US weakness than Chinese strength.[149] The comparison of China to Wal-Mart is fascinating as it positions China as a threat by equating it to one of America's greatest economic successes. The notion that Wal-Mart poses an unrepentant challenge to its competitors through ruthless business practices certainly has connections with how America views China's growing economy; giants who eliminate competition through sheer size. Thus, the association of China to Wal-Mart provides an excellent parable for America's general fears of a rapidly, and asymmetrically, growing Chinese economy.

4.5.3 The preservation of values

Critics also pointed to the erosion of American values to suggest that CNOOC's Unocal bid be disallowed. Although much more subjective than the quantifiable aspects of economic inequity or levels of resource scarcity, the impact on American values, perhaps more than anything else, helped to mobilize Congress and the American public against the sale of Unocal to CNOOC. American values were contrasted with the nature of the PRC, and even the Chinese as a people, and served to further Other China from the United States. The Communist Party and the Yellow Peril were used to illustrate China's challenge to American values.

The Chinese Communist Party and Unocal

The fact that the PRC ostensibly remains communist continues to influence American perceptions of China. The *NSS* of 2002 made specific mention that China would strive to remain communist with a government ruled by one party, and by doing so it positioned China as an adversary to America. When addressing Congress, Rep. Barton demonstrated the way in which China was being positioned as an opponent to the US when he stated that

> CNOOC is a front company for the Communist Chinese government. Seventy percent of the equity in the company is owned by the Communist Chinese government. The money that is going to be used to buy Unocal comes from the Communist government in the form of a loan.[150]

It is interesting to note how Barton attempted to contrast the US 'Self' with the Chinese 'Other' by referring to the Chinese government as 'Communist' three times in as many sentences.

Richard Pombo also took the chance to use its communist credentials against China in an attempt to malign CNOOC's Unocal bid. Referring to Unocal's sale he said "We cannot afford to have a major U.S. energy supplier controlled by the Communist Chinese.... If we allow this sale to go forward we are taking a huge risk".[151] As well, despite the efforts of Fu Chengyu to alleviate US fears as to the apolitical nature of the bid, his Party connections[J] made his protestations seem underhanded and only served to shroud the deal in more uncertainty. Joseph Khan even wrote that *The People's Daily* cited an aphorism coined by Fu declaring that "excelling at political work makes us stronger competitors".[152] Thus, even in the post-Cold War era, I argue that perceptions of Chinese Communism had a large impact on the CNOOC/Unocal affair.

Despite their different claims to governance, Thomas Friedman states that, "the Chinese and U.S. economies have become totally intertwined. While we have been focused on 9/11 and Iraq, China and America have become, in economic terms, Siamese twins".[153] The way that the US and China have reconciled their different approaches to governance and economic growth is through the 'Tiananmen-Texas bargain' which for the Chinese involves the implicit deal struck after Tiananmen between the government and the people where China's people give up their political freedom in return for a promise of 9 percent annual growth.[154] China's political stability rests completely on this compromise. The American aspect of the bargain is that it would ignore China's undervalued currency so that the US could continue to buy cheap goods and Chinese investments in the US would keep the American economy afloat. Friedman explains this delicate balance when he states that "We might see our dollar policy as a market adjustment, but they could see it as an attempt as regime change".[155] So the result of this is that "China's leaders are wary of trusting their economic growth, and perhaps the longevity of the Communist Party, to American oil companies and the Pentagon".[156] Thus, another perspective could suggest that the CCP believed the higher price it was willing to pay for Unocal was worth it in order to avoid political instability. However, to Americans, such as James Woolsey, selling Unocal to "the world's largest communist dictatorship ... [was] beyond the pale, given the nature of the Chinese government".[157]

The rearticulation of the Yellow Peril: China as the New Japan

China had by 2005, for many, become the Japan of the 1980s in its pursuit of American companies. The Chinese began to show interest in companies which were not only visibly American, but were also increasingly strategically sensitive. CNOOC's attempt at Unocal underscored the influence the ESD had on American perceptions of ES and represented the limit to which the US was willing to accommodate commercial Chinese overtures. In the 1980s the Yellow Peril re-entered the American consciousness. With its massive economic growth and threat to many US industries, despite its close political relations with the US, many Americans saw Japan as its major competitor. However, the Asian Financial Crisis in 1997 represents the handover from Japan to China as the incarnation of the Yellow

---

J Fu Chengyu was CNOOC's Party secretary and an ally of Hu Jintao.

Peril to the US. I argue that the failure of CNOOC to acquire Unocal less than a decade later is a direct consequence of this as the CTD became pervasive. Richard Siklos effectively sums up the sentiments when he states that

> Even though Unocal's American oil production represents less than 1 per cent of consumption there (and CNOOC has pledged to keep that production available domestically), parallels are already being drawn with the beginnings of the Japanese investment influx of the 1980s, perhaps best recalled by the acquisitions of the Rockefeller Center and the Empire State Building.... The 'Chinese-are-the-new-Japanese' scaremongering line is, of course, a bit bogus.... Nonetheless, even the idea of another culture having the edge on mighty America can be hysteria-inducing to its inhabitants.[158]

The Yellow Peril, in conjunction with the CTD, suggested that China's rise could fundamentally upset the traditional Western-, US-centric world order.

The last great handover in power, from Great Britain to the United States in the early-to-middle twentieth century, was unprecedented in the way in which it was peacefully achieved wherein "The two sides not only resolved various disputes and disagreements which could have provoked a conflict, but also fostered a strategic alliance which lasts to this day".[159] However, while Japan never seriously challenged the hegemony of the United States, and although China has not yet been able to equal it, serious concern was raised in the US when these powers, not only foreign, but culturally alien as well, looked to expand beyond the parameters within which the US would happily have them kept. While Japan's investment into America was very visible, albeit non-strategic, Chinese investment into America was increasingly seen to be an affront to US independence. Concurrent with CNOOC's bid for Unocal, Chinese firms Lenovo and Haier made bids for IBM's personal computer division as well as for Maytag, the company behind Hoover vacuums. These bids served to raise public awareness and concern over China's growing confidence. Heather Connon explains that

> Although Haier blames its decision to pull out of the bidding for Maytag on price, Maytag's managers may also have been reluctant to recommend the Chinese bid, with the risks of having owners whose culture they did not really understand, in preference to one from Whirlpool, from the same Anglo-Saxon stock. The cultural differences clearly increase the already considerable risks of doing cross-border deals.... It is worth remembering that the last time there was such a political storm in the US over foreign purchases of its assets, Japanese companies were in the firing line, with accusations that their property purchases in New York would mean all Manhattan would be turned Japanese.[160]

America's discomfort with growing Asian economic, and cultural, integration was increased with the Chinese' companies' bids for the American assets listed above, and reached an apogee with CNOOC's bid for Unocal.

Viewing the CNOOC bid through the optics of expansion

Another significant factor contributing to America's wariness of China is the lack of understanding by America of China. Just as 'Japan' became a reminder of US economic stagnation in the 1980s, 'China', perhaps, has become a reminder of more recent US hegemonic stagnation. Although it was argued in Chapter 2 that China is no longer as exotic as it was in the sixteenth and seventeenth centuries, there are those who argue that America still does not understand China. James McGregor writes that

> We're losing the intelligence war against China.... No, not the one with spy satellites, human operatives and electronic eavesdropping. I'm talking about intelligence: having an intelligent understanding of and intelligent discussions about China – where it's heading, why it's bidding to buy major U.S. companies and whether we should worry. Above all, I'm talking about formulating and pursuing intelligent policies for dealing with China.[161]

This lack of understanding makes otherwise benign gestures and actions, including the 'go out' strategies of the NOCs such as CNOOC, seem threatening as they appear to have sinister motives.

Returning to the theme of the Yellow Peril, parallels have been drawn between the rise of Imperial Japan in the 1930s and 1940s with contemporary China as both have required huge imports of foreign oil to fuel their growth. Japan's dependency on oil necessitated its attack on the US in the Pacific in the Second World War and "The outcry over China's potential acquisition of Unocal may or may not partake of the same kind of historical dynamic".[162] However, there is no certainty that history will repeat itself in such a fashion. In fact, if history is any indicator, China could be content to enjoy a privileged, yet nonexpansionist or aggressive role in the international system. Indeed, if CNOOC were to have won the Unocal bid it would have bought a company whose operations were largely in Asia. The expansionist notion of CNOOC's bid is challenged when one considers that "70 percent of Unocal's oil and gas reserves are in Asia, and mostly under long-term contract to Asian nations".[163] There is, therefore, strong evidence to suggest that the US fundamentally misread Chinese intentions through CNOOC's bid for Unocal and rather than being perceived as an aggressive action, the bid could have been seen as an attempt by China to meet its oil needs through the legitimate mechanisms of the international market.

Although it miscalculated its bid for Unocal, China, perhaps, understands America much better than America does China. Even though its NOCs are relatively new players, they are striving to understand the global oil market and acclimatize themselves to the accepted business practices. For instance, over two decades, CNOOC has demonstrated itself to be very Western in its approach to business, with its board meetings being conducted in English, and an American management style which was instilled by its American-educated chairman, Fu Chengyu.[164] Despite this, however, it seems that the Chinese mindfulness of its own history is echoed by CNOOC as it codenamed its bid for Unocal "Treasure Hunting Ship", an obvious nod to Zheng He.

4.6 Conclusion

Through the analysis of discourse surrounding China, CNOOC, and its bid for Unocal, some aspects relating to popular perceptions of China by America become quite clear. Although no attempt has been made in this book to quantify the number of times they have been used, several themes and adjectives relating to China recur. Rising, expanding, powerful, backwards, cunning, unfair, and communist. These have all been associated with China in one way or another and it is fascinating that although they all seem to be intrinsically linked with popular perceptions of China, they are oftentimes contradictory. This lack of understanding, or at least a *cohesive* understanding, of China is endemic to the way in which China has been positioned in opposition to the United States, and perhaps helps to clarify why Chinese actions are often perceived to be threatening. If fear stems from the unknown, this could well explain US fears of China. Chengxin Pan explores the political economy of fear which is the Western desire to uncover empirical 'truths' which allow fear to emerge as a form of knowledge, or anti-knowledge, which is a primary factor in the creation of the Other.[165] Fear is deployed in the absence of knowledge, and becomes a learned response through cultural immersion. Fear has been systematically deployed in American politics in order to bolster American interests and in so doing bolster the American Self. In essence, "To better understand the dynamics of power/knowledge/desire in the China threat paradigm, we need to recognise this paradigm for what I think it is, namely, a particular form of desire – fear – disguised as certain knowledge".[166]

Through a thorough examination of CNOOC's 2005 bid for Unocal I have hoped to demonstrate how the CTD and the ESD worked discursively with official and non-official discourses to systematically Other China to the United States. The exceptional nature of the case is highlighted by the fact that CNOOC's bid followed a decade of successful foreign investment into the US oil industry. Examples of these successful cases, many of which were significantly larger and more contentious from a financial point of view, were explored in section 4.2. The political pressure which CNOOC faced in 2005 was absent from the bids which preceded it and helps to highlight the discursive relationship between the CTD and ESD. I argue that the CNOOC/Unocal affair acted as a conduit through which other issues of the China threat theory found a unifying voice. Issues such as the United States' anger over China's currency manipulation, China's military modernization and expansion, China's relations with its neighbours, and, vitally, China's growing impact on the 'American way of life' all fall under the umbrella of matters which were raised with the CNOOC/Unocal affair. Speaking on NPR, Steve Inskeep summarized this phenomenon when he said,

> I really think this was something like a "Perfect Storm." You had people who were hostile to the bid itself, although in a way they were the smallest

minority. There are people who are just angry at China for a whole host of reasons. There are evangelical Christians, who feel China is bad to Christians. There are human rights groups who say China is bad on human rights. There are US manufacturers who say China is unfairly competing with them, and all of those groups are very active in lobbying Congress. And I think everybody who had a reason to fight China was fighting them on this one [CNOOC's bid for Unocal].[167]

By fighting China on the specific issue of CNOOC's bid for Unocal, America was able to fight China on all the others. As well, it has been suggested that the backlash to CNOOC's proposed bid was conversely given legitimacy by the other quarrels the US had with China. Crucially, it has been argued that the ESD has served to legitimize the CTD as it provided material 'evidence' and a sort of justification for fear of China. By referencing a discourse based on issues such as scarcity and market value, the China threat appears to gain a measure of empirical validity, and the CNOOC/Unocal affair epitomized this process.

Notes

1 Mortished, Carl. 2005. "China Eyes Europe After Unocal Rebuff". *The Times*, 3 August.
2 "Chinese Companies Abroad: The Dragon Tucks In". 2005. *The Economist*, 2 July.
3 Lohr, Steve. 2005. "The Big Tug of War Over Unocal". *New York Times*, 6 July.
4 Schortgen, Francis. 2006. "Protectionist Capitalists vs. Capitalist Communists: CNOOC's Failed Unocal Bid In Perspective". *Asia Pacific: Perspectives* VI(2): 2–11.
5 Hansen, Lene. 2006. *Security as Practice: Discourse Analysis and the Bosnian War*. New York: Routledge.
6 Mouawad, Jad. 2005. "Foiled Bid Stirs Worry for U.S. Oil". *New York Times*, 11 August.
7 Udall, Tom. 2005. *Understanding the Peak Oil Theory: Hearing Before the Subcommittee on Energy and Air Quality. Committee of Energy and Commerce*. Washington, DC, U.S. Government Printing Office.
8 Elliott, E. Donald. 2013. "Why the United States Does Not Have a Renewable Energy Policy". *Environmental Law Institute* 43(2): 10095–10101, p. 10097.
9 Ibid., p. 10097.
10 "Bush: 'If There Was a Magic Wand, I'd Be Waiving It'". 2008. *The Wall Street Journal*, 29 April.
11 Bush, George W. 2004. "Remarks in a Discussion at Mid-States Aluminum Corporation in Fond du Lac, Wisconsin". In *The American Presidency Project*, edited by Gerhard Peters and John T. Woolley. From www.presidency.ucsb.edu/ws/?pid=72689, 14 July 2001.
12 Obama, Barak. 2011. "Remarks at Georgetown University". In *The American Presidency Project*, edited by Gerhard Peters and John T. Woolley. From www.presidency.ucsb.edu/ws/index.php?pid=90196&st=energy+security&st1=china, 30 March 2011.
13 Roby, Martha. 2005. "159 Cong. Rec. 92, (daily ed. 25 June 2013)", *GPO.gov*, from www.gpo.gov/fdsys/granule/CREC-2013–06–25/CREC-2013–06–25-pt1-PgH4015/content-detail.html.
14 Hudson, Richard. 2005. "159 Cong. Rec. 92, (daily ed. 25 June 2013)", *GPO.gov*, from

www.gpo.gov/fdsys/granule/CREC-2013–06–25/CREC-2013–06–25-pt1-PgH4015/ content-detail.html.

15 U.S. House, Committee on Energy and Commerce. 2001. *National Energy Policy: Crude Oil and Refined Petroleum Products*. Hearing, 30 March 2001, (Serial No. 107–12). Washington, Government Printing Office, 2001.

16 Elliott. 2013. "Why the United States Does Not Have a Renewable Energy Policy", p. 10098.

17 Energy Information Administration (EIA). 2013. "Oil: Crude and Petroleum Products Explained", *Oil: Crude and Petroleum Products Explained*. Retrieved 11 December 2012, from www.eia.gov/energyexplained/index.cfm?page=oil_home#tab2.

18 Hale, Briony. 2002. "Shell Trails BP's Lead". *BBC News*. From http://news.bbc. co.uk/1/hi/business/1906963.stm, 2 April 2002.

19 EIA. 2001. "Aspects of the Refining/Marketing Joint Ventures of Shell Oil, Star Enterprises, and Texaco". *EIA Energy Finance*. From www.eia.gov/archive/emeu/ mergers/stindex.html, 23 July 2001.

20 "Shell and Texaco 'Merger Talks'". 1998. *BBC Online Network*. From http://news. bbc.co.uk/1/hi/business/127244.stm, 6 July 1998.

21 EIA. 2001. "Aspects of the Refining/Marketing Joint Ventures of Shell Oil, Star Enterprises, and Texaco".

22 "BP and Amoco in Oil Mega-Merger". 1998. *BBC Online Network*. From http:// news.bbc.co.uk/1/hi/149139.stm, 11 August 1998.

23 Moore, John Frederick. 1998. "BP to Acquire Amoco". *CNN Money*, from http:// money.cnn.com/1998/08/11/deals/bp/, 11 August 1998.

24 "BP and Amoco in Oil Mega-Merger". 1998. *BBC Online Network*.

25 Ibrahim, Youssef M. 1998. "British Petroleum Is Buying Amoco in $48.2 Billion Deal". *New York Times*, 12 August.

26 Ibid.

27 Buckingham, Lisa. 1998. "BP's £30 Billion Takeover of Amoco Approved". *Guardian*, 31 December.

28 "BP Amoco Signs Deal With FTC, Acquires ARCO". 2000. *Oil & Gas Journal*, 24 April.

29 "Green Light for BP-Arco Merger". 2000. *BBC News*. From http://news.bbc.co.uk/1/ hi/business/712962.stm, 14 April 2000.

30 Rivera Brooks, Nancy. 1999. "BP Amoco Will Acquire Arco for $27 Billion". *Los Angeles Times*, 1 April.

31 Salpukas, Agis. 1999. "It's Official: BP Is Planning To Buy ARCO". *New York Times*, 2 April.

32 Rivera Brooks. 1999. "BP Amoco Will Acquire Arco for $27 Billion".

33 Salpukas. 1999. "It's Official: BP Is Planning To Buy ARCO".

34 "Chronology: Big Oils' Years of Merger Mania". 2011. *Reuters.* From www.reuters. com/article/2011/07/14/us-conocophillips-mergers-idUSTRE76D56V20110714, 14 July 2011.

35 Nanto, Dick K., James K. Jackson, Wayne M. Morrison, and Lawrence Kumins. 2005. "China and the CNOOC bid for Unocal: Issues for Congress". *Congressional Research Service*, Washington, DC: The Library of Congress, p. 1.

36 "Why China's Unocal Bid Ran Out of Gas". 2005. *BloombergBusinessweek*. Retrieved 8 September 2013, from www.businessweek.com/stories/2005–08–03/ why-chinas-unocal-bid-ran-out-of-gas.

37 "Chinese Companies Abroad: The Dragon Tucks In". 2005. *The Economist*, 2 July.

38 Weisman, Jonathan. 2005. "In Washington, Chevron Works to Scuttle Chinese Bid". *Washington Post*, 16 July.

39 Slivinski, Stephen. 2007. "The Corporate Welfare State: How the Federal Government Subsidizes U.S. Businesses". *CATO Institute*, Washington, DC.

40 Morath, Eric. 2013. "Government's $421 Billion Bailout Turns Profitable". *The Wall Street Journal*, 21 November.
41 Mouawad, Jad. 2005. "Congress Calls for a Review of the Chinese Bid for Unocal". *New York Times*, 27 July.
42 "Chinese Companies Abroad: The Dragon Tucks In". 2005. *The Economist*. 2 July.
43 Brady, Andrew. 2013. "Walmart is a Ruthless Corporate Monster – But It's Not Too Big to Fail". *International Business Times*, 14 November.
44 Clelland, Grant. 2005. "China May Be Overpaying in Scramble for US Assets". *The Business*, 26 June.
45 Reynolds, Thomas. 2005. "Reynolds Challenges China's Bid for Unocal". *Project Vote Smart*. Retrieved 12 October 2013, from http://votesmart.org/public-statement/111630/#.UpDqMsQ73o8.
46 Davidson, Michael. 2013. "Transforming China's Energy Grid: Sustaining the Renewable Energy Push". *East Winds*. Retrieved 1 November 2013, from http://theenergycollective.com/michael-davidson/279091/transforming-china-s-grid-sustaining-renewable-energy-push.
47 Hufbaur, Gary Clyde, Yee Wong, and Ketki Sheth. 2006. *US–China Trade Disputes: Rising Tide, Rising Stakes*. Washington, DC: Peterson Institute for International Economics, August 2006, p. 50.
48 "Fu Chengyu". 2005. *The Times*. 8 July.
49 Pombo, Richard. 2005. 151 Cong. Rec. 11 (28 June 2005 to 13 July 2005), Available from *United States Government Printing Office*, Washington, 2005.
50 Marchick, David, Mark Plotkin, and David Fagan. 2005. "National Security Regulation of Foreign Investments and Acquisitions in the United States". *China Law & Practice*, June, p. 23.
51 Nanto *et al.* 2005. "China and the CNOOC bid for Unocal", p. 1.
52 Ibid.
53 Spencer, Richard. 2005. "China's US Ambitions Thwarted as Two Major Bids Scuppered". *Daily Telegraph*. 21 July.
54 Nanto *et al.* 2005. "China and the CNOOC bid for Unocal", p. 14.
55 Ney, Bob. 2005. "151 Cong. Rec. 90, (daily ed. June 30, 2005)", *GPO.gov*, from www.gpo.gov/fdsys/pkg/CREC-2005–06–30/html/CREC-2005–06–30-pt1-PgH5570–2.htm.
56 Lohr, Steve. 2005. "Who's Afraid of China Inc?" *New York Times*, 24 July.
57 Samuelson, Robert J.2005. "China's Oil Bid: A Battle to Avoid..". *Washington Post*, 6 June.
58 Lloyd-Smith, Jake. 2005. "Anti-China Rhetoric as Unocal Ponders Offers". *The Evening Standard*, 15 July.
59 National Security Strategy of the United States of America (NSS). 2002. *National Security Strategy of the United States of America*. The White House. Washington, DC.
60 Bush, George W. 2001. "Address Before a Joint Session of the Congress on the United States Response to the Terrorist Attacks of September 11". In *The American Presidency Project*, edited by Gerhard Peters and John T. Woolley. From www.presidency.ucsb.edu/ws/?pid=64731, 20 September 2001.
61 Kaplan, Robert D. 2005. "How We Would Fight China". *The Atlantic*, June.
62 Barr, Michael. 2011. *Who's Afraid of China? The Challenge of Chinese Soft Power*. London: Zed Books Ltd., p. 126.
63 Ibid.
64 Ibid.
65 Brownstein, Ronald. 2005. "Fair Play at Issue in Unocal Bid". *Los Angeles Times*, 20 July.
66 McGregor, James. 2005. "Advantage, China". *Washington Post*. 31 July.
67 NSS. 2002. *National Security Strategy of the United States of America*.

68 Ibid.
69 Ibid.
70 Lohr, Steve. 2005. "Unocal Bid Denounced at Hearing". *New York Times*, 14 July.
71 Schortgen. 2006. "Protectionist Capitalists vs. Capitalist Communists", p. 2.
72 Samuelson. 2005. "China's Oil Bid: A Battle to Avoid..".
73 Griffiths, Katherine. 2005. "Business Analysis: Chinese Assault on Unocal Raises Hackles in Energy-Obsessed US". *Independent*, 15 July.
74 NSS. 2002. *National Security Strategy of the United States of America.*
75 Dorn, James A. 1999. "Normalize Trade With China". *CATO Institute*. Retrieved 8 February 2014, from www.cato.org/publications/commentary/normalize-trade-china-0.
76 NSS. 2002. *National Security Strategy of the United States of America.*
77 Kaplan. 2005. "How We Would Fight China".
78 Cillizza, Chris. 2013. "The 113th Congress? More Partisan Than the 112th Congress – Thanks to Republicans". *Washington Post*, 26 September.
79 Greidinger, Marc. 1991. "The Exon-Florio Amendment: A Solution in Search of a Problem". *American University International Law Review* 6(2): 111–177, p. 113.
80 Ibid.
81 Barboza, David and Andrew Ross Sorkin. 2005. "Chinese Oil Company Offers $18.5 Billion for Unocal". *New York Times*, 22 June.
82 Lohr, Steve. 2005. "Unocal Bid Opens Up New Issues of Security". *New York Times*, 13 July.
83 Marchick *et al.* 2005. "National Security Regulation of Foreign Investments and Acquisitions in the United States".
84 Lohr. 2005. "Unocal Bid Opens Up New Issues of Security".
85 "Chinese Oil Firm Drops Unocal Bid". 2005. *Guardian*, 2 August.
86 Litterick, David. 2005. "China Angered by US Oil Sale Intervention, Beijing Calls on Washington to Stop 'Interfering' in Free Enterprise". *Daily Telegraph*, 5 July.
87 Lee, Don and Elizabeth Douglass. 2005. "Chinese Drop Takeover Bid for Unocal". *Los Angeles Times*, 3 August.
88 "H.Res.344 – Expressing the sense of the House of Representatives that a Chinese state-owned energy company exercising control of critical United States energy infrastructure and energy production capacity could take action that would threaten to impair the national security of the United States". (H.Res.344, 30 June 2005) U.S. Congress, 109 (2005), Available at GPO.gov, www.gpo.gov/fdsys/pkg/BILLS-109hres344ih/html/BILLS-109hres344ih.htm.
89 Hufbaur *et al.* 2006. *US–China Trade Disputes*, p. 47.
90 U.S. House, Committee on Finance. 1996. *China Most-Favored-Nation (MFN) Status.* Hearing, 6 June 1996 (S. Hrg. 104–871). Washington: Government Printing Office, 1997.
91 Lohr. 2005. "Unocal Bid Opens Up New Issues of Security".
92 Kilpatrick, Carolyn. 2005. "151 Cong. Rec. 11 (28 June 2005 to 13 July 2005)", *United States Government Printing Office*, Washington, 2005.
93 Bartholomew, Carolyn. 2005. *Dark Clouds on the Horizon: The CNOOC-Unocal Controversy and Rising U.S.-China Frictions*, Carnegie Endowment for International Peace, Washington, DC, 14 July 2005.
94 Wolfers, Arnold. 1952. " 'National Security' as an Ambiguous Symbol". *Political Science Quarterly* 67(4): 481–502, p. 481.
95 Ibid.
96 Nanto *et al.* 2005. "China and the CNOOC bid for Unocal", p. 2.
97 Barboza and Sorkin. 2005. "Chinese Oil Company Offers $18.5 Billion for Unocal".
98 Calder, Kent E. 1996. *Asia's Deadly Triangle: How Arms, Energy and Growth Threaten to Destabilize Asia Pacific*. London: Nicholas Brealey.
99 Wayne, Leslie and David Barboza. 2005. "Unocal Deal: A Lot More Than Money is At Issue". *New York Times*, 24 June.

100 Lohr. 2005. "The Big Tug of War Over Unocal".
101 Jaffe, Amy Myers. 2005. "Wasted Energy". *New York Times*, 27 July.
102 Lohr. 2005. "Unocal Bid Opens Up New Issues of Security".
103 Taylor, Jerry. 2005. "CNOOC Bid for Unocal No Threat to Energy Security". *CATO Institute*, from www.cato.org/publications/free-trade-bulletin/cnooc-bid-unocal-no-threat-energy-security, 19 July 2005.
104 Young, Alf. 2005. "US Would be Wise to Oil Wheels of China's Progress". *The Herald*. 28 June.
105 IEA. 2012. *Oil & Gas Security: Emergency Response of IEA Countries*. Paris: OECD/IEA: 1–19.
106 O'Rourke, Ronald. 2013. "China Naval Modernization: Implications for U.S. Navy Capabilities – Background and Issues for Congress". *Congressional Research Service*. Washington, DC: 1–115, p. 2.
107 Ibid., p. 41.
108 Blustein. 2005. "Many Oil Experts Unconcerned Over China Unocal Bid".
109 O'Rourke. 2013. "China Naval Modernization".
110 Erickson, Andrew S. 2012. "China's Modernization of Its Naval and Air Power Capabilities". In *Strategic Asia 2012–13: China's Military Challenge*, edited by Ashley J. Tellis and Travis Tanner. Seattle: National Bureau of Asian Research.
111 Wines, Michael and Edward Wong. 2011. "China's Push to Modernize Military Is Bearing Fruit". *New York Times*, 5 January.
112 Cody, Edward. 2005. "Unocal Bid Shows China Needs Oil for Growth". *Washington Post*, 30 June.
113 Ibid.
114 Carpenter, Ted Galen. 2013. "Tightrope Diplomacy: U.S. Arms Sales to Taiwan". *CATO Institute*. From www.cato.org/publications/commentary/tightrope-diplomacy-us-arms-sales-taiwan, 31 October 2013.
115 Havemann, Joel and Elizabeth Douglass. 2005. "Lawmakers Seek to Stop CNOOC Bid". *Los Angeles Times*, 14 July.
116 Blustein, Paul. 2005. "Many Oil Experts Unconcerned Over China Unocal Bid". *Washington Post*, 1 July.
117 Barrionuevo, Alexei. 2005. "Foreign Suitors Nothing New in Oil Patch". *New York Times*, 1 July.
118 Pelosi, Nancy. 2005. "Pelosi Statement on Amendment to Block Chinese Bid to Acquire Unocal". *Pelosi.house.gov*. From http://pelosi.house.gov/news/press-releases/pelosi-statement-on-amendment-to-block-chinese-bid-to-acquire-unocal, 30 June 2005.
119 Khan, Joseph. 2005. "A Deft Balance in Orchestrating China's Oil Offer". *New York Times*, 7 July.
120 Wayne and Barboza. 2005. "Unocal Deal: A Lot More Than Money is At Issue".
121 Bartholomew. 2005. *Dark Clouds on the Horizon*.
122 Jaffe. 2005. "Wasted Energy".
123 Ibid.
124 Blumenthal, Dan. 2005. "Providing Arms: China and the Middle East". *Middle East Quarterly* 12(2): 11–19.
125 Morrison, Wayne M. 2005. "China–U.S. Trade Issues". *Congressional Research Service*. Washington, DC, 19 July.
126 Lohr. 2005. "The Big Tug of War Over Unocal".
127 Ibid.
128 Wayne and Barboza. 2005. "Unocal Deal: A Lot More Than Money is At Issue".
129 Brownstein. 2005. "Fair Play at Issue in Unocal Bid".
130 Barton, Joe. 2005. "151 Cong. Rec. 90, (daily ed. 30 June 2005)", *GPO.gov*, from www.gpo.gov/fdsys/pkg/CREC-2005-06-30/html/CREC-2005-06-30-pt1-PgH5570-2.htm.

131 Nanto *et al.* 2005. "China and the CNOOC bid for Unocal".
132 Ibid., p. 9.
133 Worstall, Tim. 2011. "Chuck Schumer and the China Currency Bill". *Forbes*, 16 October.
134 Andrews, Edmund. 2005. "Capitol Nearly Speechless on Big China Bid". *New York Times*, 24 June.
135 Tschang, Chi-Chu. 2008. "What Bud's Takeover Means for China". *Bloomberg-Businessweek*, 14 July.
136 Morrison. 2005. "China-U.S. Trade Issues".
137 Barrionuevo, Alexei and Andrew Ross Sorkin. 2005. "Chevron Criticizes Rival Suitor". *New York Times*, 25 June.
138 Lohr. 2005. "The Big Tug of War Over Unocal".
139 Litterick. 2005. "China Angered by US Oil Sale Intervention, Beijing Calls on Washington to Stop 'Interfering' in Free Enterprise".
140 Jefferson, William. 2005. "151 Cong. Rec. 90, (daily ed. June 30, 2005)", *GPO.gov*, from www.gpo.gov/fdsys/pkg/CREC-2005–06–30/html/CREC-2005–06–30-pt1-PgH5570–2.htm.
141 Ikensen, Daniel J. 2009. "Hard Lessons from the Auto Bailouts". *CATO Institute*. Retrieved 7 February 2014, from www.cato.org/policy-report/novemberdecember-2009/hard-lessons-auto-bailouts.
142 Wayne and Barboza. 2005. "Unocal Deal: A Lot More Than Money is At Issue".
143 Bayh, Evan. 2005. "151 Cong. Rec. 106 (daily ed. July 29, 2005)", *GPO.gov*, from www.gpo.gov/fdsys/pkg/CREC-2005–07–29/html/CREC-2005–07–29-pt1-PgS9436.htm.
144 Siklos, Richard. 2005. "Sell to the Chinese if Their Money is Green". *Sunday Telegraph*, 10 July.
145 "H.Res.344 – Expressing the sense of the House of Representatives that a Chinese state-owned energy company exercising control of critical United States energy infrastructure and energy production capacity could take action that would threaten to impair the national security of the United States". (H.Res.344, 30 June 2005) U.S. Congress, 109 (2005), Available at GPO.gov, www.gpo.gov/fdsys/pkg/BILLS-109hres344ih/html/BILLS-109hres344ih.htm.
146 Barrionuevo. 2005. "Foreign Suitors Nothing New in Oil Patch".
147 Lohr. 2005. "Who's Afraid of China Inc?"
148 World Bank. 2014. "China Overview". *The World Bank*. Retrieved 28 May 2014, from, www.worldbank.org/en/country/china/overview.
149 "Chinese Strength, US Weakness". 2005. *New York Times*, 3 August.
150 Barton, Joe. 2005. "151 Cong. Rec. 90, (daily ed. 30 June 2005)", *GPO.gov*, from www.gpo.gov/fdsys/pkg/CREC-2005–06–30/html/CREC-2005–06–30-pt1-PgH5570–2.htm.
151 Barrionuevo. 2005. "Foreign Suitors Nothing New in Oil Patch".
152 Khan. 2005. "A Deft Balance in Orchestrating China's Oil Offer".
153 Friedman, Thomas L. 2005. "Joined at the Hip". *New York Times*, 20 July.
154 Ibid.
155 Ibid.
156 Khan, Joseph. 2005. "Behind China's Bid for Unocal: A Costly Quest for Energy Control". *New York Times*, 24 June.
157 Lohr. 2005. "Unocal Bid Denounced at Hearing".
158 Siklos. 2005. "Sell to the Chinese if Their Money is Green".
159 Yongping, Feng. 2006. "The Peaceful Transition of Power from the UK to the US". *Chinese Journal of International Politics* 1: 83–108.
160 Connon, Heather. 2005. "China Learns to Scale the Wall: But a Hunger for Deals Can be Hampered by Differences in Commercial Culture". *Observer*, 24 July.
161 McGregor. 2005. "Advantage, China".

162 Jaffe. 2005. "Wasted Energy".
163 Lohr. 2005. "The Big Tug of War Over Unocal".
164 Lee and Douglass. 2005. "Chinese Drop Takeover Bid for Unocal".
165 Pan, Chengxin. 2012. *Knowledge, Desire and Power in Global Politics: Western Representations of China's Rise*. Cheltenham, UK: Edward Elgar Publishing Limited.
166 Ibid., p. 67.
167 Davidson, Adam. 2005. "Chinese Oil Company Pulls Unocal Bid". *NPR*, Retrieved 12 October 2013, from www.npr.org/templates/story/story.php?storyId=4782101.

5 Conclusion

5.1 Summing up

In its broadest sense, my goal in this book was to challenge everyday assumptions about Sino-American relations and to understand how China has been positioned as a threat to the US. In order to accomplish this, I explored themes of the China threat and energy security within broader American discourse to see how China has been positioned as a contender to US interests. Significantly, I avoided the sometimes popular assumption that China poses an a priori threat to the US and also eschewed the notion that China's rise will inevitably lead it into conflict with the United States. Despite the fact that future conflict between the US and China should not be presupposed, China does stand out as a legitimate contender to US dominance and its re-emergence will undoubtedly impact greatly on US foreign policy choices. The purpose of this book was to examine US/Western readings of China's re-emergence to show how foreign policy choices are shaped by, and themselves contribute to, central China Threat and Energy Security Discourses, and specifically to examine how the ESD has contributed to the entrenchment of the CTD.

Due to the fact that explorations of energy security in relation to China's rise are not new to the literature I sought to make my own contribution to this body of work and this was predicated on three goals. First, I examined the themes of the CT as well as ES in order to provide clear and detailed accounts of what they are and how they are both products of, and instigators of, particular popular Western perceptions. Through these examinations I defined the central discourses of the CTD and the ESD. It was argued that in-depth analysis of these two central discourses was necessary because invocation in the literature of these themes, while extensive, has tended to remain ambiguous because the meaning of the CT and ES are rarely fully articulated by those who invoke them. By studying these themes in isolation to see how they had been deployed in broader American discourse I aimed to redress this trend. Exploring the CT and ES in isolation allowed me to undertake the second contribution of the book, which was to offer a contextual analysis of how the CTD and the ESD have been mobilized together. This contextual analysis offered insight into how the discursive relationship which exists between these two central discourses becomes actualized. Explorations into the CTD are augmented by contextual analysis and I used issues of ES to illustrate the import of China's re-emergence and to emphasize the role of the CTD in US conceptions of modern China. Although it is just a component part of the CTD, contemporary ES concerns have helped to focus other CT perceptions so that they have coalesced into one relatively organized discourse, and the CNOOC/Unocal affair provides an extremely potent example of how this resulted in practice. The final goal of the book was a comprehensive analysis of CNOOC's bid for Unocal in 2005 as it represents a defining, yet previously overlooked event which has helped to shape contemporary Sino-American relations. I aimed to show how poststructuralism can extend beyond the narrow confines of IR theory to be used in practice. Moving away from strictly theoretical debates I used poststructuralism to analyse substantive policy-relevant issues and to do so I undertook an in-depth analysis of CNOOC's failed 2005 bid for Unocal. Although this event represents a defining moment in contemporary Sino-American relations and served to shape relations in the years following 2005, it had not yet been addressed in detail, and this book serves to remedy this oversight. The examination of the case study helped to demonstrate how the event continues to affect relations between the two powers, especially with regard to ES, a decade after it took place.

This book will conclude by briefly addressing the CTD, the ESD, and will outline the legacy of CNOOC's bid for Unocal. This will allow us to reflect on what we have learnt from the case study and extrapolate these lessons to broader issues beyond this book's focus of analysis. The conclusion will therefore demonstrate how CNOOC's Unocal bid provides us with a platform from which we can move on to other analyses. Importantly, I will explain how the way poststructuralism was used in the analysis of the case study can be used to examine other important contemporary issues.

5.2 The virtue of poststructuralism

I did not aim to make a substantive contribution to IR theory but poststructuralism was discussed as it was essential to the analysis in the book. In order to examine the central discourses and to explore the mutually reinforcing relationship which exists between them, poststructural discourse analysis was used, and the approach was heavily indebted to that which Hansen devised in *Security as Practice*. Encapsulating prevalent attitudes in IR, Christian Reus-Smit writes, "It is now commonplace to bemoan our field's lack of relevance, and to blame this sorry situation on our penchant for ever-more abstract theorising over the analysis of real-world problems".[1] Certainly, there is an evident lack of coherence, or "list of essential differences", within the discipline which some have taken to suggest implies significant fracture within IR.[2] While traditionalists within the field may be derisive of increasingly critical approaches, and while those outside the field may even be critical of the discipline on the whole, through its analysis of the CNOOC/Unocal affair this book has aimed to demonstrate the very real value of critical postpositivist approaches to 'real world' analysis. McKenna highlights the relationship between critical theory and practice and explains that "critical studies must avowedly deal with the lived experiences of our times".[3] I have attempted to demonstrate that taking a critical stance does not entail highlighting the shortcomings of prior research without offering an alternative, and that adopting a poststructural approach does not necessarily deprive a study of a lack of structure and doom it to incoherence. Specifically, I have sought to demonstrate that far from being an academic and theoretical self-indulgence, poststructuralism actually offers the best method for exploring and analysing issues of the 'real world' because it can engage with the everyday elements and incremental factors which are ignored by positivist theories with rigid structures but which are essential to a proper understanding of practice. Thus, the growing support which poststructuralism has gained within the discipline must be maintained.

This theoretical predilection has clear impact on policy-relevant studies. Chengxin Pan writes, "Among the most reported stories in the first decade of the twenty-first century, topping the list was not the global financial crisis, the long-running Iraq War, or even the 'September 11' terrorist attacks – it was the rise of China".[4] Although much work has been done on China's rise this book has argued that the CTD has become entrenched in Western perceptions of China at the expense of competing discourses. This has privileged particular readings of China in which it is positioned as an antagonist to US interests. Despite positivist attempts at the empirical exploration of objective truths, understanding China in this manner is doomed: "*Contra* positivism, we cannot bypass thoughts and representations to come into direct contact with China as it is. What we see as 'China' cannot be detached from various discourses and representations of it".[5] The exploration into positivist-based why-questions surrounding China's rise are, to return to Doty's language, "incomplete in an important sense".[6] The book abjures the positivism which is engendered by many orthodox IR approaches, including realism and thin variants of constructivism, and argues that freedom from the dogmatic assumptions of causality allows for much fuller analytic latitude. Rather than participating in IR approaches which lay claim to universalist theories of cause and effect, poststructuralism allows us to look beyond the why-questions of causal claims of social action to the how-questions of discursive construction.[7] Eschewing the problems associated with presuppositions of subjectivity, linked as they are to problematic ideas of intentions, behaviour, and the motivating factors behind them, allows us to examine the production of meanings.[8] Without an understanding of how popular Western perceptions of China have emerged, there is little hope that we can examine these perceptions in a meaningful way.

Because poststructuralism is concerned with the production of knowledge, and because what constitutes knowledge is not universal, poststructural approaches are therefore themselves not universal. While I used Hansen's analysis of the Bosnian War as a model for this book, the requisite and significant changes necessary to make a poststructural approach applicable to an analysis of the CNOOC/Unocal affair were made. The fact that such an approach cannot simply be grafted from one study to another means that a certain amount of finesse and contextual modelling is required to make it relevant to something new and this differs from positivists' assertions that theory should be universal. However, despite the effort which is necessary to make a poststructural approach relevant to analyses I maintain that due to its specificity poststructuralism can actually tell us much more about a particular subject than can any positivist approach.

My aim was that by using poststructuralist discourse analysis to engage with actual policy-relevant analyses, poststructuralism would be vindicated as a method for 'real world' analysis. I also sought to demonstrate the degree and nature of the insight we can expect from poststructuralism. Most basically, while it allows for very detailed understanding of the themes and how they interact in a case study, poststructuralism, does not gift us any predictive powers. While I argue this is true of all IR approaches, poststructuralism does not feign to offer these powers while positivist theories do. Although it is possible for us to speculate about future Chinese action, it is impossible for us to read Chinese intentions with any degree of certainty, and IR approaches which claim to offer such insight are dangerously misleading. However, predilection towards this notion of forecasting is not uncommon in positivist-dominated IR. These trends can be evidenced in the discourse which surrounds CNOOC's bid for Unocal as US elites often misinterpreted (perhaps intentionally at times) Chinese actions as aggressive and this served to further alienate the Chinese Other from the US Self. As it assiduously avoided analysis of intentions, the book did not aim to provide any prediction about future US-Sino relations or to provide any prescriptive account of how to mitigate perceptions of the growing threat China poses to the US. In fact, because the analysis placed emphasis on discursive rather than causal elements to examine the CT and ES, this approach avoided any pretence that it offers predictive power – a promise common to more orthodox theoretical approaches. Discourse analysis does not offer any predictive power, nor does it attempt to. Indeed, an analysis of CNOOC's unsuccessful bid for Unocal would not have predicted the adroitness CNOOC has demonstrated in successfully buying into the North American oil industry in the years since 2005. CNOOC's $15.1 billion purchase of Nexen is perhaps most illustrative of this.[A] Through its analysis, however, what the book suggested is that better and more thorough understandings of the way the CT has been discursively constructed will allow for better understandings of wider Sino-American relations. Through its analysis of discourse surrounding China threat and dominant energy security themes, this book aimed to illustrate how the CT has actually been constructed and how it is representative of particular readings of China's 'rise' and how it is based on particular perceptions rather than any 'fact'.

Thus, while the utility of a poststructural approach might not be immediately appealing to key industry insiders or oil company executives who might prefer

A See: Rocha, Euan. 2013. "CNOOC Closes $15.1 Billion Acquisition of Canada's Nexen". *Reuters*, 25 February.

to explore cause-and-effect through empirical analysis of data, I make the argument that such analysis would be incomplete without the ability to look behind the numbers to find out what they *mean*. Examining how issues of identity affect foreign policy choices are indicative of this. Although the application of post-structuralism to the analysis of a project requires certain delicacy, the analytical output offers the sophistication to show us, for instance, how the CTD and ESD can augment our understanding of growth trends in Chinese energy use.

While Chinese competition, growth, and oil acquisition (as a primary tenet of the ESD) had been positioned as separate issues which impacted directly on US national security, it was demonstrated that the pressure CNOOC faced in its Unocal bid was made possible by the interaction of the CTD and ESD. If one or the other of the basic discourses had not been mobilized, the outcome of CNOOC's bid for Unocal could well have been different, and examples of successful foreign acquisitions of US oil companies were provided in order to support this. It was demonstrated that while there was a general trend of anti-China sentiment in official and non-official US discourse, the issue of ES, and the CNOOC/Unocal affair in particular, helped to focus this somewhat fractured sentiment onto one particular issue and in so doing helped to amplify and legitimize the CTD. The poststructural discourse analysis which was used offered unique insight into *how* such sentiment was mobilized by the US elites and *how* it was accepted by the US audience.

I also sought to demonstrate the importance of incremental factors (e.g. negative Chinese stereotypes in American literature) and changes (e.g. growing demand for oil within China) which are key to different discourses. The ability to explore these discursive factors is essential to an in-depth understanding of the international system and changes within it. The importance the analysis in this book placed on these elements, which can be trivialized by other approaches, helped to demonstrate some shortcomings of securitization theory and this helped to show that the Western/US relationship with China is more complex than a binary between 'China as threat' and 'China as opportunity'. While the Copenhagen School made invaluable contributions to critical security studies, including an emphasis on the importance of speech-acts in creating politically rather than analytically motivated security concerns, securitization is undermined by its formulaic approach to these issues. Its clear divisions between securitization and desecuritization, as well as its conception of security as an 'exceptional' concern which is elevated above the plane of 'normal' politics, rests on a set of agreed-upon rules. The analysis of the discursive relationship between the two discourses highlighted problematic aspects which resulted from attempts to make a poststructural approach to security concerns commensurate with the rules of securitization; the examination of the CTD and the ESD, as well as the analysis of CNOOC's bid for Unocal, provided evidence that served to further challenge the notion of securitization. While the book argues that the general trend described by securitization occurred as China was actively positioned as a threat to US interests, the inability for securitization theory to account for the incremental and very nuanced – but very real – everyday Sino-American tensions ultimately means that China cannot be said to have been securitized.

5.3 Why does CNOOC's bid matter?

CNOOC's rebuff by American elites and the US audience in its bid for Unocal was indicative of relations between China and the United States in 2005 and the event is particularly representative of their energy relations at that point. However, the significance of the case extends beyond this narrow scope. Despite its definite location and the fact that it is a product of a specific time and place I argue that there are lessons we can glean from it which go beyond its specific temporal and spatial context. I believe these lessons are specific to the case study itself, but can also be understood more generally from the way the case study was analysed. In this last section I will explore this latter aspect first, suggesting how discourse analysis could be used to explore broad issues of Self/Other dynamics involving US relations, as well as more specific issues of the impact of ES on these relations. I will then explore specific lessons which have emerged from CNOOC's Unocal bid.

The CNOOC/Unocal affair as a template: transplanting the research method

While it does not offer us a theoretical framework within which we can extrapolate future American or Chinese actions, poststructuralism offers us great insight into how American conceptions of China as a competitor have been constructed, and understanding this process can give us insight into relations on a wider level. Using a similar approach to analysis could give us detailed readings into other constitutive elements of Sino-American relations and the discourse which creates them. I feel that there are two obvious avenues of exploration to this end. First, although ES is a primary tenet of the CTD, many other issues reinforce CT perceptions in US/Western discourse and these perceptions could also provide fertile ground for analysis. Thus, more detailed explorations of how cultural artefacts, for example, affect discursive relations between the US and China would be valuable. Moreover, although I have explained that there are pro-China or China-neutral discourses which challenge the dominance of the CTD, I have not explored them in detail. Thus, comprehensive explorations of how these perceptions exist at the margins of broader Western discourse which serves to vilify China would be highly illuminating as a counterpoint to this book. Second, understanding how the US Self has constructed China as an Other can also give us insight into how it may construct different Others with varying degrees of Otherness. Examinations could take place at a country-specific or regional level and could include competitors as well as allies. For instance, we could learn a great deal more about the 'special relationship' between the US and the UK through detailed discourse analysis. Similar examinations could also take place on a wider level to examine relations between the US and the EU. To return to a notion of competition which was raised in the introduction, poststructural discourse analysis would also be hugely beneficial in examinations of complex links between the West/US and BRICs, for example. Such examinations could be especially illuminating in the wake of Russia's annexation of Crimea and the ensuing crisis in Ukraine.

As I demonstrated in this book, such analyses would benefit from context, and as ES is an unavoidable tenet of contemporary perceptions of national security, analyses of how country- and region-specific ES discourses impact on relations with the West and with the United States would be extremely fruitful. Volatility remains a defining feature of perceptions of ES, particularly with regard to oil, and this unpredictability has a serious impact on relations between energy importers and exporters. In fact, the oil market is so unpredictable that those who are defined as importers and exporters are subject to radical change. The recent fracking boom in the United States is indicative of this as the US is now, through new technology and economic feasibility, able to tap a domestic supply of oil and gas which was hitherto unavailable and which has allowed it to act increasingly independently from those it has traditionally imported energy from. This has palpable knock-on effects on its relations with traditional exporter states, which, as was demonstrated in previous chapters, tend to be unstable. By harnessing its own domestic supplies the US is able to reduce its dependency on volatile exporters in the Middle East and Russia, and this allows it to be more assertive in these bilateral relationships. The fact that the US and fellow Western countries were able to forgo more orthodox assaults on Russia and have instead deployed their own 'gas weapon' against the Russian Federation to combat its increasing aggression in Eastern Europe is indicative of how changeable the constitutive elements of ES can be. With this new-found security, Sino-American relations, and concerns surrounding 'red oil' in the US could change as well. To address this we will now return to CNOOC's 2005 bid.

The CNOOC/Unocal affair as a defining event

Despite the way the method of analysis can be expanded to include other cases and issues, there are specific lessons which should be taken from the exploration of the book's case study itself. The most significant conclusion stems from the fact that the CT is most effectively understood as a discursive construction. Thus, the CT is neither 'true' nor 'false' but as a dominant reading of China in the West it has a direct impact on foreign policy. This reading, however, must not be understood as a universal 'fact', but rather as a contextually dependent perception which is prone to change. This change means that while it is the defining discourse now, we should not necessarily expect that the CTD will represent Western perceptions of China in perpetuity, and notions of a 'strong' China need not be solely associated with a menacing China.

Whereas China was often portrayed as being weak and ineffectual in the precursor discourses, the CTD stands alone in that it portrays China as an increasingly strong and assertive Other. Much like the way in which the Balkanization discourses based the Otherness of the Balkans through a "temporal move, which constitutes this uncivilized violent, hateful, and barbaric Balkan identity as 'ancient'", the pre-China Threat Discourses similarly temporally situate China in the past and ascribe to it attributes of 'underdevelopment' and 'backward-ness'.[9] The CTD, on the other hand, firmly places China in the present and situates it as an economic, military, and even ideological contender to the US, which made CNOOC's bid for Unocal incredibly unpalatable to CTD proponents unfamiliar with a modern, rising China. Robert Kaplan highlights the threat theory: "Pulsing with consumer and martial energy ... China constitutes the principal conventional threat to America's liberal imperium".[10]

ES, like the CT, is a defining feature of contemporary international politics which is best understood through discourse analysis. Essential as it is to international politics, this book placed added emphasis on ES through its thorough examination of the CNOOC/Unocal affair. Although ES, like China's rise, is predicated on temporal notions of progress, unlike the pre-China Threat Discourses, the ESD is a relatively recent phenomenon. Its arrival, by happenstance, has coincided with the advent of the CTD. A result of this concurrence is that while an analysis of the ESD or the CDT may take place in isolation, the China threat can best be understood alongside an exploration of a particular ES strategy, and CNOOC's bid for Unocal is an extremely fertile source for such exploration. The symbiosis between these two basic discourses is well explained by a segment from the *New York Times* written during the height of the CNOOC/Unocal affair, which states:

> When analysts and economic historians look back, this summer may well prove to be the turning point in Chinese-American relations, the time when America chose short-range paranoia over rational behavior. From the dozen or so proposals in Congress for across-the-board tariffs against Chinese imports to the Pentagon's rumblings about Chinese military buildups, the rhetoric from Washington keeps escalating. America seems to be on the run, fuelled by the false perception that China's rapid economic rise poses an inevitable threat to the United States. By repeatedly demonizing China, Washington risks creating the hostility it fears.[11]

This is the crux of the China threat. The US perceives China as a threat while it simultaneously perceives itself to be increasingly incapable of stemming the threat should China wish to act against American interests. The perception of American energy insecurity is a potent indicator of feelings that America can no longer bully China about.[12] The Unocal bid represents the transition from Chinese accession to Western/US interests to a new period where China would no longer capitulate to pressure in this way. It therefore also represents the emergence of a more assertive China.

Thus, I argue that the CNOOC/Unocal affair stands out as an event which focused and defined the China threat, and as such it is intrinsic to the CTD and demonstrates perfectly how the ESD has served to reinforce CT sentiments within Western discourse. Just as perceptions of a weak and backwards China started to shift with Deng's 'opening up' in the late-1970s, and how perceptions of China-as-adversary re-materialized with the Tiananmen crackdown, I argue that the backlash to CNOOC's bid for Unocal in 2005 signals the emergence of the modern China threat. Although the CTD was dominant at the time, the bid signalled the end of a period in which the US/West was able to corral Chinese ambitions which were perceived to be against Western interests. China has not been similarly stymied in the years since the summer of 2005 and perceptions of Chinese capabilities have now been twinned with perceptions of Chinese threats, and this is evidenced by CNOOC's success with Nexen. Therefore, CNOOC's bid for Unocal stands alone as an event where the two central discourses interacted. Thus, CNOOC's bid not only provides us with a case study which illustrates how discourse operates to affect relations and policy, but it stands alone as a milestone in Sino-American relations.

Notes

1 Reus-Smit, Chrisian. (2012). "International Relations, Irrelevant? Don't Blame Theory". *Millennium – Journal of International Studies* 40: 525–540, 525.
2 Lawson, George. (2012). "The Eternal Divide? History and International Relations". *European Journal of International Relations* 18: 203–226, 204.
3 McKenna, Bernard. (2004). "Critical Discourse Studies: Where To From Here?" *Critical Discourse Studies* 1(1): 9–39, 27.
4 Pan, Chengxin. (2012). *Knowledge, Desire and Power in Global Politics: Western Representations of China's Rise*. Cheltenham, UK: Edward Elgar Publishing Limited. p. vii.
5 Pan. (2012). *Knowledge, Desire and Power in Global Politics*. p. vii.
6 Doty, Roxanne Lynn. (1993). "Foreign Policy as Social Construction: A Post-Positivist Analysis of U.S. Counterinsurgency Policy in the Philippines". International Studies Quarterly, 37(3): 297–320, 298.
7 Wendt, Alexander. (1987). "The Agent-Structure Problem in International Relations Theory". *International Organization* 41(3): 335–370.
8 Doty, Roxanne Lynn. (1993). "Foreign Policy as Social Construction: A Post-Positivist Analysis of U.S. Counterinsurgency Policy in the Philippines". *International Studies Quarterly* 37(3): 297–320.
9 Hansen, Lene. (2006). *Security as Practice: Discourse Analysis and the Bosnian War*. New York: Routledge. p. 107.
10 Kaplan, Robert D. (2005). "How We Would Fight China". *The Atlantic*. Washington, DC, Jay Lauf. June.
11 "America's Summer of Discontent". (2005). *New York Times*. 11 August 2005.
12 Andrews, Edmund. (2005). "Capitol Nearly Speechless on Big China Bid". *New York Times*. 24 June 2005.

Appendix 1
House Resolution 344–109th Congress, 1st Session

Whereas oil and natural gas resources are strategic assets critical to national security and the Nation's economic prosperity; Whereas the global demand for oil and natural gas is at the highest levels in history; Whereas the global excess capacity of oil production, at between 1,500,000 and 2,000,000 barrels per day, is at its lowest level in the past several decades, contributing to world oil prices reaching historic highs of above $60 per barrel; Whereas natural gas globally is the fastest growing component of primary energy consumption, projected to increase by nearly 70 percent by 2025; Whereas the National Security Strategy of the United States approved by President George W. Bush on 17 September 2002, concludes that the People's Republic of China remains strongly committed to national one-party rule by the Communist Party; Whereas China's daily consumption of crude oil grew by nearly 850,000 barrels in 2004, accounting for more than one-third of the increase in world demand for oil in 2004; Whereas China's consumption of crude oil is expected to grow by an additional 7.5 percent in 2005, and world oil prices are projected to rise significantly as a result of increasing demand from China for oil; Whereas notwithstanding the increasing demand from China for oil, domestic Chinese output of oil has remained relatively stagnant; Whereas on 23 June 2005, the China National Offshore Oil Corporation (CNOOC) announced its intent to acquire Unocal Corporation, in the face of a competing bid for Unocal Corporation from Chevron Corporation; Whereas the People's Republic of China owns approximately 70 percent of CNOOC; Whereas a significant proportion of the CNOOC acquisition is to be financed and heavily subsidized by banks owned by the People's Republic of China; Whereas Unocal Corporation is based in the United States, and has approximately 1,750,000,000 barrels of oil equivalent, with its core operating areas in Southeast Asia, Alaska, Canada, and the lower 48 States; Whereas CNOOC has made various representations about its intent to sell oil developed in the Gulf of Mexico to the United States, but has not made any commitments to sell other natural gas and oil it develops into global energy markets instead of shipping it directly to China; Whereas a CNOOC acquisition of Unocal Corporation would result in the strategic assets of Unocal Corporation being preferentially allocated to China by the Chinese government; Whereas a Chinese government acquisition of Unocal Corporation would weaken the ability of the

United States to influence the oil and gas supplies of the Nation through companies that must adhere to United States law; Whereas Unocal Corporation was responsible for the production of energy equivalent to approximately 411,000 barrels of oil per day in 2004, which is approximately one-third of all global excess oil production capacity; Whereas the petroleum sector uses a range of sensitive technologies for exploration (such as seismic analysis and processing, downhole logging sensors, and modelling software), including technologies that have "dual-use" commercial and military applications; Whereas several of the technologies used in oil and energy production require export licensing for export from the United States to China; Whereas the CNOOC acquisition of Unocal Corporation could provide access to Unocal Corporations sensitive dual-use technologies that the United States would otherwise restrict for export to China; Whereas oil companies owned by the People's Republic of China are active in parts of the world, such as Sudan and Iran, that are subject to United States sanctions laws, and the national security of the United States is threatened by the export of sensitive, export controlled, and dual-use technologies to such countries; Whereas barriers to the ability of the United States Government to enforce export controls and sanctions could pose a direct threat to the national security of the United States; and Whereas section 721 of the Defense Production Act of 1950 (50 App. U.S.C. 2170) authorizes the President to suspend or prohibit any foreign acquisition, merger, or takeover of a United States corporation that threatens the national security of the United States, if the President finds that "there is credible evidence that leads the President to believe that the foreign interest exercising control might take action that threatens to impair the national security" and other provisions of law "do not in the President's judgement provide adequate and appropriate authority for the President to protect the national security": Now, therefore, be it

Resolved, That it is in the sense of the House of Representatives that –

(1) the Chinese state-owned China National Offshore Oil Corporation, through control of Unocal Corporation obtained by the proposed acquisition, merger, or takeover of Unocal Corporation, could take action that would threaten to impair the national security of the United States; and

(2) if Unocal Corporation enters into an agreement of acquisition, merger, or takeover of Unocal Corporation by the China National Offshore Oil Corporation, the President should initiate immediately a thorough review of the proposed acquisition, merger, or takeover (H5571, 30–06–2005).

Bibliography

Abraham, Spencer. 2001. "Meet Energy Challenges; Start Providing Security at Home". *Washington Times*, 29 November.

Alston, Margaret. 2013. "Gender Mainstreaming and Climate Change". *Women's Studies International Forum* 47: 287–294.

Ambrosio, Thomas. 2012. "The Rise of the 'China Model' and 'Beijing Consensus': Evidence of Authoritarian Diffusion?" *Contemporary Politics* 18(4): 381–399.

"America's Summer of Discontent". 2005. *New York Times*, 11 August 2005.

Andrews, Edmund L. 2007. "Candidates Offer Different Views on Energy Policy". *New York Times*, 28 November.

Andrews, Edmund. 2005. "Capitol Nearly Speechless on Big China Bid". *New York Times*, 24 June.

Angelis-Dimakis, Athanasios, Marcus Biberacher, Javier Dominguez, Guilia Fiorese, Sabine Gadocha, Edgard Gnansounou, Giorgio Guariso, Avraam Kartaldis, Luis Pan-ichelli, Irene Pinedo, and Michella Roba. 2011. "Methods and Tools to Evaluate the Availability of Renewable Energy Sources". *Renewable and Sustainable Energy Reviews* 15: 1182–1200.

Arora-Jonsson, Seema. 2011. "Virtue and Vulnerability: Discourses on Women, Gender and Climate Change". *Global Environmental Change* 21: 744–751.

Ashley, Richard and RBJ Walker. (1990). "Reading Dissidence/Writing the Discipline: Crisis and the Question of Sovereignty in International Studies". *International Studies Quarterly* 34(3): 367–416.

Asia Pacific Energy Research Centre (APERC). 2007. *A Quest for Energy Security in the 21st Century*. Japan: Institute of Energy Economics.

Avraham, Eli and Anat First. 2006. "'I Buy American': The American Image as Reflected in Israeli Advertising". *Journal of Communication* 53(2): 282–299.

Bailey, Paul. 2001. *China in the Twentieth Century*. Oxford: Blackwell Publishers Ltd.

Barboza, David and Andrew Ross Sorkin. 2005. "Chinese Oil Company Offers $18.5 Billion for Unocal". *New York Times*, 22 June.

Bardi, Ugo. 2013. "The Grand Challenge of the Energy Transition". *Frontiers in Energy Research* 1: 1–4.

Barr, Michael. 2011. *Who's Afraid of China? The Challenge of Chinese Soft Power*. London: Zed Books Ltd.

Barton, Joe. 2005. "151 Cong. Rec. 90, (daily ed. 30 June 2005)", *GPO.gov*, from www. gpo.gov/fdsys/pkg/CREC-2005-06-30/html/CREC-2005-06-30-pt1-PgH5570-2.htm.

Bates, Gill. 1998. "Chinese Military Modernization and Arms Proliferation in the Asia-Pacific". In *In China's Shadow: Regional Perspectives on Chinese Foreign Policy and*

Military Development, edited by Jonathan D. Pollack and Richard H. Yang. Washington, DC: RAND Corporation, 10–36.

Bartholomew, Carolyn. 2005. *Dark Clouds on the Horizon: The CNOOC-Unocal Controversy and Rising U.S.–China Frictions*, Carnegie Endowment for International Peace, Washington, DC, 14 July 2005.

Barrionuevo, Alexei. 2005. "Foreign Suitors Nothing New in Oil Patch". *New York Times*, 1 July.

Barrionuevo, Alexei and Andrew Ross Sorkin. 2005. "Chevron Criticizes Rival Suitor". *New York Times*, 25 June.

Bayh, Evan. 2005. "151 Cong. Rec. 106 (daily ed. July 29, 2005)", *GPO.gov*, from www.gpo.gov/fdsys/pkg/CREC-2005-07-29/html/CREC-2005-07-29-pt1-PgS9436.htm.

Bekar, Clifford and Richard G. Lipsey. 2002. *Science, Institutions and the Industrial Revolution*. Burnaby: Simon Fraser University.

Beresford, Charles (Lord Charles William De la Poer Beresford [1st Baron]). 1899. *The Break-up of China, With an Account of its Present Commerce, Currency, Waterways, Armies, Railways, Politics and Future Prospects*. London: Harper & Brothers.

Bernkopf Tucker, Nancy. 2012. *The China Threat: Memories, Myths, and Realities in the 1950s*. New York: Columbia University Press.

Blumenthal, Dan. 2005. "Providing Arms: China and the Middle East". *Middle East Quarterly* 12(2): 11–19.

Blustein, Paul. 2005. "Many Oil Experts Unconcerned Over China Unocal Bid". *Washington Post*, 1 July.

Blustein, Paul and Mike Musgrove. 2005. "U.S. May Scrutinize IBM's China Deal". *Washington Post*, 25 January.

Bonnett, Alastair. 2004. *The Idea of the West: Culture, Politics and History*. Basingstoke: Palgrave Macmillan.

Booth, Ken. 1991. "Security and Emancipation". *Review of International Studies*, 17(4): 313–326.

Booth, Ken. (2007). *Theory of World Security*. Leiden: Cambridge University Press.

BP. 2011. "Statistical Review of World Energy 2011". *BP.com*. Retrieved 24 February 2012, from www.bp.com/sectionbodycopy.do?categoryId=7500&contentId=7068481.

BP. 2012. "The BP Energy Outlook 2030". *BP.com*. Retrieved 24 February 2012, from www.bp.com/genericarticle.do?categoryId=9003467&contentId=7067432.

BP. 2014. "BP Energy Outlook 2035". *BP.com/energy outlook*, Retrieved 16 February 2014, from www.bp.com/content/dam/bp/pdf/Energy-economics/Energy-Outlook/Energy_Outlook_2035_booklet.pdf.

BP. 2014. "Review By Energy Type: Hydroelectricity". *BP.com*. Retrieved 5 October 2013, from www.bp.com/en/global/corporate/about-bp/energy-economics/statistical-review-of-world-energy-2013/review-by-energy-type/hydroelectricity.html.

"BP Amoco Signs Deal With FTC, Acquires ARCO". 2000. *Oil & Gas Journal*, 24 April.

"BP and Amoco in Oil Mega-Merger". 1998. *BBC Online Network*. From http://news.bbc.co.uk/1/hi/149139.stm, 11 August 1998.

Brady, Andrew. 2013. "Walmart is a Ruthless Corporate Monster – But It's Not Too Big to Fail". *International Business Times*, 14 November.

Brady, David and Denise Kall. 2008. "Nearly Universal, but Somewhat Distinct: The Feminization of Poverty in Affluent Western Democracies, 1969–2000". *Social Science Research* 37: 976–1007.

Brandt, Adam R. 2007. "Testing Hubbert". *Energy Policy* 35(2007): 3074–3088.

Brookings Institution. 2014. "About the Energy Security Initiative". [Online] *Energy*

Security Initiative. Retrieved 14 November 2013, from www.brookings.edu/about/projects/energy-security/about.

Broomfield, Emma V. 2003. "Perceptions of Danger: The China Threat Theory". *Journal of Contemporary China* 12(35): 265–284.

Brown, Harold, Joseph W. Prueher, and Adam Segal 2003. "Chinese Military Power: Report of an Independent Task Force Sponsored by the Council on Foreign Relations Maurice R. Greenberg Ceter for Geoeconomic Studies". *Council on Foreign Relations*. New York, Council on Foreign Relations.

Brown, Matthew H., Christie Rewey, and Troy Gagliano. (2003). *Energy Security*. Boulder, CO: National Conference of State Legislatures.

Brown, Stephen P.A. and Hillard G. Huntington. 2008. "Energy Security and Climate Change Protection: Complementarity or Tradeoff?" *Energy Policy* 36(9): 3510–3513.

Brownstein, Ronald. 2005. "Fair Play at Issue in Unocal Bid". *Los Angeles Times*, 20 July.

Bryant Jr., Keith L. 1988. *Encyclopedia of American Business History and Biography: Railroads in the Age of Regulation, 1900–1980*. New York: Facts on File.

Bryce, Robert. 2010. "Five Myths About Green Energy". *Washington Post*, 25 April.

Buckingham, Lisa. 1998. "BP's £30 Billion Takeover of Amoco Approved". *Guardian*, 31 December.

Bush, George. 1990. "The President's News Conference on the Persian Gulf Crisis". [Online] In *The American Presidency Project*, edited by Gerhard Peters and John T. Woolley. From www.presidency.ucsb.edu/ws/index.php?pid=18792&st=oil&st1=iraq, 30 August 1990.

Bush, George. 1989. "Statement on the 45th Anniversary of D-Day". In *The American Presidency Project*, edited by Gerhard Peters and John T. Woolley. From www.presidency.ucsb.edu/ws/?pid=17117, 6 June 1989.

Bush, George W. 2001. "Address Before a Joint Session of the Congress on the United States Response to the Terrorist Attacks of September 11". In *The American Presidency Project*, edited by Gerhard Peters and John T. Woolley. From www.presidency.ucsb.edu/ws/?pid=64731, 20 September 2001.

Bush, George W. 2008. "Bush Remarks on Climate". *Washington Post*, 16 April.

Bush, George W. 2008. "Bush Holds Press Conference, Addresses the Economy". *Washington Post*, 29 April.

Bush, George W. 2001. "Remarks Announcing the Energy Plan in St. Paul, Minnesota". [Online] In *The American Presidency Project*, edited by Gerhard Peters and John T. Woolley. From www.presidency.ucsb.edu/ws/index.php?pid=45617&st=energy+security&st1=oil, 17 May 2001.

Bush, George W. 2004. "Remarks in a Discussion at Mid-States Aluminum Corporation in Fond du Lac, Wisconsin". In *The American Presidency Project*, edited by Gerhard Peters and John T. Woolley. From www.presidency.ucsb.edu/ws/?pid=72689, 14 July 2001.

Bush, George W. 2008. "Remarks on Energy and Climate Change". In The American Presidency Project, edited by Gerhard Peters and John T. Woolley. From www.presidency.ucsb.edu/ws/index.php?pid=76957&st=international+cooperation+and+technology+investment&st1=.

"Bush: 'If There Was a Magic Wand, I'd Be Waiving It' ". 2008. *The Wall Street Journal*, 29 April.

Buzan, Barry. 1983. *People, States, and Fear: The National Security Problem in International Relations*. Chapel Hill: University of North Carolina Press.

Buzan, Barry and Lene Hansen. (2009). *The Evolution of International Security Studies*. Cambridge: Cambridge University Press.

Buzan Barry and Ole Waever. 2003. *Regions and Powers: The Structure of International Security*. Cambridge: Cambridge University Press.

Buzan, Barry and Ole Waever. (1997). "Slippery? Contradictory? Sociologically Untenable? The Copenhagen School Replies". *Review of International Studies*, 23: 241–250.

Buzan, Barry, Ole Waever, and Jaap de Wilde. 1998. *Security: A New Framework for Analysis*. Boulder, Colorado: Lynne Rienner Publishers, Inc.

Buzan, Barry. (1996). "The Timeless Wisdom of Realism?" In Steve Smith, Ken Booth, and Marysia Zalewski. (eds.), *International Theory: Positivism and Beyond*. Cambridge: Cambridge University Press: 47–65.

Calder, Kent E. 1996. *Asia's Deadly Triangle: How Arms, Energy and Growth Threaten to Destabilize Asia Pacific*. London: Nicholas Brealey.

Campbell, Colin J. 2006. "Understanding Peak Oil". [Online] *About Peak Oil*, Retrieved 2 August 2013, from www.peakoil.net/about-peak-oil.

Campbell, Colin J. 1988. *The Coming Oil Crisis*, Multi-Science Publishing Company & Petroconsultants S.A.

Campbell, David. (1999). "Apartheid Cartography: The Political Anthropology and Spatial Effects of International Diplomacy in Bosnia". *Political Geography*, 18: 395–435.

Campbell, David. 2013. "Poststructuralism". In *International Relations Theories: Discipline and Diversity (3rd Edition)*, edited by Tim Dunne, Milja Kurki, and Steve Smith. New York: Oxford University Press, 223–246.

Campbell, David. (1998). *Writing Security: United States Foreign Policy and the Politics of Identity*. Minneapolis: University of Minnesota Press.

Cappiello, Dina and Matthew Daly. 2012. "Republicans, Democrats at Odds on Energy Issues". *The Associated Press-NORC Center for Public Affairs Research*. From www. apnorc.org/news-media/Pages/News+Media/republicans-democrats-at-odds-on-energy-issues.aspx, 14 June 2012.

Carpenter, Ted Galen. 2013. "China's Military Spending: No Cause for Panic". [Online] *CATO Institute*. Retrieved 9 February 2014, from www.cato.org/publications/commentary/chinas-military-spending-no-cause-panic, 4 April 2013.

Carpenter, Ted Galen. 1998. "Let Taiwan Defend Itself". *CATO Policy Analysis*. From www.cato.org/pubs/pas/pa-313.html, 24 August 1998.

Carpenter, Ted Galen. 2013. "Tightrope Diplomacy: U.S. Arms Sales to Taiwan". *CATO Institute*. From www.cato.org/publications/commentary/tightrope-diplomacy-us-arms-sales-taiwan, 31 October 2013.

Carpenter, Ted Galen and James A. Dorn. 2000. "China: Constructive Partner or Emerging Threat?" *CATO Institute*. From www.cato.org/publications/commentary/china-constructive-partner-or-emerging-threat, 10 May 2000.

Carpenter, Ted Galen and Justin Logan. 2008. "Relations with China, India, and Russia". In *Cato Handbook for Policymakers: 7th Edition*, edited by David Boaz. Washington, DC: Cato Institute, 549–559.

Carter, Jimmy. 1977. "Address to the Nation on Energy". [Online] In *The American Presidency Project*, edited by Gerhard Peters and John T. Woolley. From www.presidency. ucsb.edu/ws/index.php?pid=7369&st=reduce+demand+through+conservation&st1=.

Carter, Jimmy. 1977. "The Environment Message to the Congress". In *The American Presidency Project*, edited by Gerhard Peters and John T. Woolley. From www.

presidency.ucsb.edu/ws/index.php?pid=7561&st=As+our+nation+increasingly+turns+t
o+coal+as+a+replacement+for+our+dwindling+supplies+of+oil+and+gas%2C+we+m
ust+be+sure+that+we+will+not+fall+short+of+the+goals+we+have+established+to+pr
otect+human+health+and+the+general+environment&st1=.

Carter, Jimmy. 1979. "Kansas City, Missouri Remarks at a Reception for Business and Civic Leaders". [Online] In *The American Presidency Project*, edited by Gerhard Peters and John T. Woolley. From www.presidency.ucsb.edu/ws/index.php?pid=31533 &st=cheap+oil&st1=, 15 October 1979.

Checchi, Arianna, Arno Behrens, and Christian Egenhofer. 2009. "Long-Term Energy Security Risks for Europe: A Sector-Specific Approach". *Centre for European Policy Studies* 309: 1–47.

Cheng, Dean. 2013. "Countering China's A2/AD Challenge". [Online] *The Heritage Foundation*. Retrieved 22 March 2014, from www.heritage.org/research/commen-tary/2013/9/countering-chinas-a2-ad-challenge, 20 September 2013.

"China's Military Rise: The Dragon's New Teeth". 2012. *The Economist*. 7 April.

"China's Threats". 2000. *Washington Post*. 23 February.

"Chinese Companies Abroad: The Dragon Tucks In". 2005. *The Economist*, 2 July.

"Chinese Oil Firm Drops Unocal Bid". 2005. *Guardian*, 2 August.

"Chinese Strength, US Weakness". 2005. *New York Times*, 3 August.

"Chronology: Big Oils' Years of Merger Mania". 2011. *Reuters*. From www.reuters.com/article/2011/07/14/us-conocophillips-mergers-idUSTRE76D56V20110714, 14 July 2011.

Ciarreta, Aitor, Maria Paz Espinosa, and Cristina Pizarro-Irizar. 2014. "Is Green Energy Expensive? Empirical Evidence from the Spanish Electricity Market". *Energy Policy* 69: 205–215.

Cillizza, Chris. 2013. "The 113th Congress? More Partisan Than the 112th Congress – Thanks to Republicans". *Washington Post*, 26 September.

Ciuta, Felix. 2010. "Conceptual Notes on Energy Security: Total or Banal Security?" *Security Dialogue* 41(123): 21.

Ciuta, Felix. 2009. "Security and the Problem of Context: A Hermeneutical Critique of Securitisation Theory". *Review of International Studies* 35: 301–326.

Clegg, Jenny. 1994. *Fu Manchu and the Yellow Peril: The Making of a Racist Myth*. Oakhill, Stoke-on-Trent, Staffordshire: Trentham Books Limited.

Clelland, Grant. 2005. "China May Be Overpaying in Scramble for US Assets". *The Business*, 26 June.

Clinton, William J. 2000. "Statement on the Organization of Petroleum Exporting Countries Production Decision and the Legislative Agenda for Energy Security". [Online] In *The American Presidency Project*, edited by Gerhard Peters and John T. Woolley. From www.presidency.ucsb.edu/ws/index.php?pid=58301&st=energy+security&st1=o il, 28 March 2000.

Clinton, William J. 1995. "Statement on Petroleum Imports and Energy Security". [Online] *In The American Presidency Project*, edited by Gerhard Peters and John T. Woolley. From www.presidency.ucsb.edu/ws/index.php?pid=50988&st=energy+securi ty&st1=oil, 16 February 1995.

Cody, Edward. 2005. "Unocal Bid Shows China Needs Oil for Growth". *Washington Post*, 30 June.

Cole, Robert J. 1989. "Japanese Buy New York Cachet With Deal for Rockefeller Center". *New York Times*. 31 October.

Connon, Heather. 2005. "China Learns to Scale the Wall: But a Hunger for Deals Can be Hampered by Differences in Commercial Culture". *Observer*, 24 July.

Conrad, Sebastian. 2012. "Enlightenment in Global History: A Historiographic Critique". *The American Historical Review* 117(4): 999–1027.

Cordesman, Anthony H., Ashley Hess, and Nicholas S. Yarosh. 2013. *Chinese Military Modernization and Force Development: A Western Perspective*. A Report for the CSIS Burke Chair in Strategy. Washington, DC, Center for Strategic & International Studies (CSIS).

Crane, Keith, Roger Cliff, Evan Medeiros, James Mulvenon, and William Overholt. 2005. *Modernizing China's Military: Opportunities and Constraints*. Santa Monica, CA: RAND Corporation.

Crozier, Justin. 2002. "A Unique Experiment". *China in Focus Magazine*, Society for Anglo-Chinese Understanding (SACU).

Dalby, Simon. 2002. *Environmental Security*. Minneapolis: University of Minnesota Press.

Dannreuther, Roland. 2010. "Energy Security". In *The Routledge Handbook of New Security Studies*, edited by J. Peter Burgess. London: Routledge, 144–154.

Dannreuther, Roland. 2010. *International Relations Theories: Energy, Minerals and Conflict*. Polinares: EU Policy on Natural Resources, Polinares.

Davidson, Adam. 2005. "Chinese Oil Company Pulls Unocal Bid". *NPR*, Retrieved 12 October 2013, from www.npr.org/templates/story/story.php?storyId=4782101.

Davidson, Michael. 2013. "Transforming China's Energy Grid: Sustaining the Renewable Energy Push". *East Winds*. Retrieved 1 November 2013, from http://theenergycollective.com/michael-davidson/279091/transforming-china-s-grid-sustaining-renewable-energy-push.

De Buitrago, Sybille Reinke. 2012. *Portraying the Other in International Relations: Cases of Othering, Their Dynamics and the Potential for Transformation*. Newcastle upon Tyne: Cambridge Scholars Publishing.

Dent, Christopher M. 2011. "China, Africa and Conceptualising Development Relations". In *China and Africa Development Relations*, edited by Christopher Dent. New York: Routledge, 165–178.

Dent, Christopher M. 2012. "Renewable Energy and East Asia's New Developmentalism: Towards a Low Carbon Future?" *The Pacific Review* 25(5): 561–587.

Dent, Christopher M. and Elspeth Thomson. 2013. "Asia's and Europe's Energy Policy Challenges: Introduction". *Asia Europe Journal* 11(3): 201–2010.

Der Derian, James. (1987). *On Diplomacy: A Genealogy of Western Estrangement*. Oxford: Blackwell.

Der Derian, James. (1990). "The (S)pace of International Relations: Simulation, Surveillance, and Speed". *International Studies Quarterly* 34(3): 295–310.

Derrida, Jaques. 2001. *Writing and Difference*. London and New York: Routledge Classics.

Dikotter, Frank. 1992. *The Discourse of Race in Modern China*. Stanford: Stanford University Press.

Department of Defense (DOD). 2002. "Annual Report on the Military Power of the People's Republic of China". *Report to Congress Pursuant to the FY2000 National Defense Authorization Act*. Washington, DC, United States Department of Defense: 1–56.

Department of Defense (DOD). 2012. "Annual Report to Congress: Military and Security Developments Involving the People's Republic of China 2012". *A Report to Congress Pursuant to the National Defense Authorization Act for Fiscal Year 2000*. Washington, DC, Office of the Secretary of Defense: 1–43.

Department of Defense (DOD). 2013. "Annual Report to Congress: Military and Security Developments Involving the People's Republic of China 2013". *A Report to Congress Pursuant to the National Defense Authorization Act for Fiscal Year 2000*. Washington, DC, Office of the Secretary of Defense.

Doolittle, Justus. 2000. "Chinese Parsimony". In *Sinophiles and Sinophobes: Western Views of China*, edited by Colin Mackerras. Oxford: Oxford University Press.

Dorn, James A. 1999. "Normalize Trade With China". *CATO Institute*. Retrieved 8 February 2014, from www.cato.org/publications/commentary/normalize-trade-china-0.

Doty, Roxanne Lynne. 2000. "Desire All the Way Down". *Review of International Studies* 26: 137–139.

Doty, Roxanne Lynn. 1993. "Foreign Policy as Social Construction: A Post-Positivist Analysis of U.S. Counterinsurgency Policy in the Philippines". *International Studies Quarterly* 37(3): 297–320.

Downs, Erica. 2004. "The Chinese Energy Security Debate". *The China Quarterly* 177: 21–41.

Downs, Erica. 2010. "Who's Afraid of China's National Oil Companies?" In *Energy Security: Economics, Politics, Strategies and Implications*, edited by Carlos Pascual and Jonathan Elkind. Washington, DC: Brookings.

Drogin, Bob and Eric Lichtblau. 2000. "Reno, Freeh Insist Wen Ho Lee Posed 'Great Risk' to U.S". *Los Angeles Times*. 27 September.

Du Halde, Jean Baptiste. 1776. *The General History of China: Containing A Geographical, Historical, Chronological, Political, and Physical Description of the Empire of China*. London: John Watts.

Easton, Ian. 2014. "China's Deceptively Weak (and Dangerous) Military". [Online] *The Diplomat*, Retrieved 17 March 2014, from http://thediplomat.com/2014/01/chinas-deceptively-weak-and-dangerous-military/, 31 January 2014.

Eilperin, Juliet. 2010. "Emissions Limits, Greater Fuel Efficiency for Cars, Light Trucks Made Official". *Washington Post*, 2 April.

Elliott, E. Donald. 2013. "Why the United States Does Not Have a Renewable Energy Policy". *Environmental Law Institute* 43(2): 10095–10101.

EIA. 2001. "Aspects of the Refining/Marketing Joint Ventures of Shell Oil, Star Enterprises, and Texaco". *EIA Energy Finance*. From www.eia.gov/archive/emeu/mergers/stindex.html, 23 July 2001.

Energy Information Administration (EIA). 2013. "Oil: Crude and Petroleum Products Explained", *Oil: Crude and Petroleum Products Explained*. Retrieved 11 December 2012, from www.eia.gov/energyexplained/index.cfm?page=oil_home#tab2.

Energy Information Administration (EIA). (2013b). "U.S. Imports from Iran of Crude Oil and Petroleum Products". [Online] *Petroleum & Other Liquids*. from www.eia.gov/dnav/pet/hist/LeafHandler.ashx?n=PET&s=MTTIM_NUS-NIR_2&f=A.

Environmental Protection Agency (EPA). 2013. "Clean Energy". *United States Environmental Protection Agency*. From www.epa.gov/cleanenergy/energy-and-you/affect/natural-gas.html, 25 September 2013.

Environmental Protection Agency (EPA). (2014). "Pollution Prevention (P2)" [Online] *United States Environmental Protection Agency*, from www.epa.gov/p2/.

Erickson, Andrew S. 2012. "China's Modernization of Its Naval and Air Power Capabilities". In *Strategic Asia 2012–13: China's Military Challenge*, edited by Ashley J. Tellis and Travis Tanner. Seattle: National Bureau of Asian Research.

Eriksson, Johan. 1999. "Observers or Advocates?: On the Political Role of Security Analysts". *Cooperation and Conflict* 34(3): 311–330.

Etzioni, Amati. 2011. "Is China a Responsible Stakeholder?" *International Affairs*, 87(3): 539–553.

"Expert: China Resembles the USSR Right Before the Fall". 2012. [Online] *Business Insider*. Retrieved 18 March 2014, from www.businessinsider.com/china-expert-no-one-wants-to-be-the-gorbachev-of-china-2012-6, 24 June 2012.

Fagan, Mary. 2000. "Sheikh Yamani Predicts Price Crash as Age of Oil Ends". *Telegraph*. 25 June.

Fagiani, Riccardo, Julian Barquin, and Rudi Hakvoort. 2013. "Risk-Based Assessment of the Cost-Efficiency and the Effectivity of Renewable Energy Support Schemes: Certificate Markets Versus Feed-In Tariffs". *Energy Policy* 55: 648–661.

Fairbank, John King. 1983. *The Cambridge History of China*. Cambridge: Cambridge University Press.

Fairbank, John King and Merle Goldman. 1998. *China: A New History*. Cambridge: Massachusetts, The Belknap Press of Harvard University Press.

Fass, Josef. 1973. "Chinese Newspapers". In *Essays on the Sources for Chinese History*, edited by Donald D. Leslie, Colin Mackerras, and Wang Gungwu. Columbia, South Carolina: University of South Carolina Press.

Fernald, John G. and Oliver D. Babson. 1999. "Why Has China Survived the Asian Crisis So Well? What Risks Remain?" *International Finance Discussion Papers* Federal Reserve System. 633: 1–34.

Fontana, Michela. 2011. *Matteo Ricci: A Jesuit in the Ming Court*. Plymouth: Rowman & Littlefield Publishers, Inc.

Ford, Gerald R. 1977. "Annual Message to the Congress: The Economic Report of the President". [Online] In *The American Presidency Project*, edited by Gerhard Peters and John T. Woolley. From www.presidency.ucsb.edu/ws/index.php?pid=5570&st=oil&st 1=energy+security, 18 January 1977.

Foucault, Michel. 1972. *The Archaeology of Knowledge and the Discourse on Language*. New York: Pantheon Books.

Foucault, Michel. 2002. *Power*. London: Penguin.

Foucault, Michel. 1980. *Power/Knowledge: Selected Interviews and Other Writings, 1972–1977*. Brighton: Harvester Press.

Frank, Caroline. 2011. *Objectifying China: Chinese Commodities in Early America*. Chicago: The University of Chicago Press.

Friedman, Thomas L. 2003. "A War for Oil?" *New York Times*. 5 January.

Friedman, Thomas L. 2005. "Joined at the Hip". *New York Times*, 20 July.

"Fu Chengyu". 2005. *The Times*, 8 July.

Fuerth, Leon. 2005. "Energy, Homeland, and National Security". In *Energy and Security: Toward a New Foreign Policy Strategy*, edited by Jan. H. Kalicki and David L. Goldwyn. Baltimore: The Johns Hopkins University Press.

Gaffney, Frank Jr. 2005. "Gaffney Gives Hill Testemony on Unocal". [Online] *Center for Security Policy*. Retrieved 9 March 2014, from www.centerforsecuritypolicy. org/2005/07/13/gaffney-gives-hill-testimony-on-unocal-2/.

Galbraith, Kate. 2013. "Deep-Sea Drilling Muddies Political Waters". *New York Times*. 6 February.

Gaonkar, Dilip Paremeshwar. 2002. "Toward New Imaginaries: An Introduction". *Public Culture* 14(1): 1–19.

"George Osborne: China 'An Opportunity, not a threat'". 2013. [Online] *BBC News*. Retrieved 27 March 2015, from www.bbc.co.uk/news/uk-politics-24512532, 13 October 2013.

Gibson, James L. 1988. "Political Intolerance and Political Repression During the McCarthy Red Scare". *The American Political Science Review* 82(2): 511–529.

Gjorv, Gunhild Hoogensen. 2012. "Security by Any Other Name: Negative Security, Positive Security, and a Multi-Actor Security Approach". *Review of International Studies* 38(4): 835–859.

Glaser, Bonnie S. and Brittany Billingsley. 2011. "Is China's Aircraft Carrier a Threat to U.S. Interests?" *Center for Strategic & International Studies (CSIS)*. From http://csis. org/publication/chinas-aircraft-carrier-threat-us-interests, 11 August 2011.

Gompert, David C. 2013. *Sea Power and American Interests in the Western Pacific*. Santa Monica, CA: RAND Corporation.

"Green Light for BP-Arco Merger". 2000. *BBC News*. From http://news.bbc.co.uk/1/hi/ business/712962.stm, 14 April 2000.

Greidinger, Marc. 1991. "The Exon-Florio Amendment: A Solution in Search of a Problem". *American University International Law Review* 6(2): 111–177.

Griffiths, Katherine. 2005. "Business Analysis: Chinese Assault on Unocal Raises Hackles in Energy-Obsessed US". *Independent*. July 15.

Hale, Briony. 2002. "Shell Trails BP's Lead". *BBC News*. From http://news.bbc.co.uk/1/ hi/business/1906963.stm, 02 April 2002.

Halper, Stefan. 2011. "The China Threat". Foreign Policy. Retrieved 08 February 2014, from www.foreignpolicy.com/articles/2011/02/22/the_china_threat, 22 February 2011.

Hansen, Lene. 2006. *Security as Practice: Discourse Analysis and the Bosnian War*. New York: Routledge.

Hargreaves, Steve. 2011. "Oil's Future in Deepwater Drilling". *CNN Money*. From http:// money.cnn.com/2011/01/11/news/economy/oil_drilling_deepwater/, 11 January 2011.

Harrison, Matthew S. and Kecia M. Thomas. 2009. "The Hidden Predjudice in Selection: A Research Investigation on Skin Color Bias". *Journal of Applied Social Psychology* 39(1): 134–168.

Havemann, Joel and Elizabeth Douglass. 2005. "Lawmakers Seek to Stop CNOOC Bid". *The Los Angeles Times*, July 14.

Hayter, Susan. 2004. *The Social Dimension of Global Production Systems: A Review of the Issues*. Working Paper No. 25. Geneva, International Labour Organization: 1–25.

Helm, Dieter. 2011. "Peak Oil and Energy Price – A Critique". *Oxford Review of Economic Policy* 27(1): 68–91.

Hirshberg, Matthew S. 1993. "Consistency and Change in American Perceptions of China". *Political Behavior* 15(3): 247–263.

Hoft, Jim. 2013. "Study: Weak Sissy Men More Likely to Support Welfare State, Wealth Redistribution, Democrats". *Gateway Pundit*. Retrieved 5 October 2013, from www. thegatewaypundit.com/2013/05/study-weak-sissy-men-more-likely-to-support-welfare-state-wealth-redistribution-democrats/.

"How Gas Price Controls Sparked '70s Shortages". 2006. *Washington Times*. 15 May.

"H.Res.344 – Expressing the sense of the House of Representatives that a Chinese state-owned energy company exercising control of critical United States energy infrastructure and energy production capacity could take action that would threaten to impair the national security of the United States". (H.Res.344, 30 June, 2005) U.S. Congress, 109 (2005), Available at *GPO.gov*, from www.gpo.gov/fdsys/pkg/BILLS-109hres344ih/ html/BILLS-109hres344ih.htm.

Hsu, Hsuan L. 2012. "Sitting in Darkness: Mark Twain and America's Asia". *American Literary History* 25(1): 69–84.

Huang, Yukon. 2012. "China's Rise: Opportunity or Threat for East Asia?", [Online] *The*

Carnegie Endowment for International Peace, Retrieved 27 March 2015, from http://carnegieendowment.org/ieb/2012/04/12/china-s-rise-opportunity-or-threat-for-east-asia, 12 April 2012.

Hubbert, M. King. 1962. *Energy Resources: A Report to the Committee on Natural Resources of the National Acadamy of Sciences – National Research Council*. Washington, DC: National Acadamy of Sciences – National Research Council.

Hudson, Richard. 2005. "159 Cong. Rec. 92, (daily ed. 25 June 2013)", *GPO.gov*, from www.gpo.gov/fdsys/granule/CREC-2013-06-25/CREC-2013-06-25-pt1-PgH4015/content-detail.html.

Hufbaur, Gary Clyde, Yee Wong, and Ketki Sheth. 2006. *US-China Trade Disputes: Rising Tide, Rising Stakes*. Washington, DC: Peterson Institute for International Economics, August 2006.

Humphreys, Jasper. 2012. "Resource Wars: Searching for a New Definition". *International Affairs* 88(5): 1065–1082.

Huysmans, Jef. 2011. "What's in an Act? On Security Speech Acts and Little Security Nothings". *Security Dialogue* 42: 371–383.

Ibrahim, Youssef M. 1998. "British Petroleum Is Buying Amoco in $48.2 Billion Deal". *New York Times*, 12 August.

Iijima, Chris K. 2001. "When the 'Model Minority' Meets Fu Manchu". *Honolulu Star-Bulletin*, 2 May.

Ikenson, Daniel J. 2006. "China: Mega-Threat or Quiet Dragon". [Online] *CATO Institute*. From www.cato.org/publications/speeches/china-megathreat-or-quiet-dragon, 6 March 2006.

Ikensen, Daniel J. 2009. "Hard Lessons from the Auto Bailouts". [Online] *CATO Institute*. Retrieved 7 February 2014, from www.cato.org/policy-report/novemberdecember-2009/hard-lessons-auto-bailouts.

Intergovernmental Panel on Climate Change (IPCC). 2000. *IPCC Special Report: Emissions Scenarios – Summary for Policymakers*. Intergovernmental Panel on Climate Change, Intergovernmental Panel on Climate Change: 1–21.

International Energy Agency (IEA). 2013. *2013 Key World Energy Statistics*. Paris: OECD/IEA.

International Energy Agency (IEA). 2013. "Energy Efficiency". [Online] *International Energy Agency*. Retrieved 4 October 2013, from www.iea.org/topics/energyefficiency/.

International Energy Agency (IEA). 2013. *Resources to Reserves 2013: Oil, Gas and Coal Technologies for the Energy Markets of the Future*. Paris: OECD/IEA.

International Energy Agency (IEA). 2013. "Topic: Renewables". [Online] *International Energy Agency*. Retrieved 4 October 2013, from www.iea.org/topics/renewables/.

International Energy Agency (IEA). 2012. *Oil & Gas Security: Emergency Response of IEA Countries*. Paris: OECD/IEA.

International Institute for Strategic Studies (IISS). 2013. "China's Defense Spending: New Questions". *International Institute for Strategic Studies*. From www.iiss.org/en/publications/strategic%20comments/sections/2013-a8b5/china-39-s-defence-spending-new-questions-e625, 2 August 2013.

International Renewable Energy Agency (IRENA). 2014. "About IRENA". *International Renewable Energy Agency (IRENA)*. From www.irena.org/Menu/index.aspx?PriMenuID=13&mnu=Pri.

Isenberg, David. 2011. "Pentagon Talks Up China Threat". [Online] *CATO Institute*. Retrieved 8 February 2014, from www.cato.org/publications/commentary/pentagon-talks-china-threat, 30 August 2011.

Jaffe, Amy Myers. 2005. "Wasted Energy". *New York Times*, 27 July.

Jansen, Klaus. 2013. "German Anxieties Over China's Rise". *Deutsche Welle*. 20 August. www.dw.de/german-anxieties-over-chinas-rise/a-16963665.

Jarvis, Darryl S. L. 1999. *International Relations and the Challenge of Postmodernism: Defending the Discipline*. Columbia, SC: University of South Carolina Press.

Jefferson, William. 2005. "151 Cong. Rec. 90, (daily ed. 30 June 2005)", *GPO.gov*, from www.gpo.gov/fdsys/pkg/CREC-2005-06-30/html/CREC-2005-06-30-pt1-PgH5570-2.htm.

Jevons, W. Stanley. 1865. *The Coal Question: An Inquest Concerning the Progress of the Nation, and the Probable Exhaustion of Our Coal-Mines*. London and Cambridge: Macmillan and Co.

Ji, You. 2007. "Dealing With the Malacca Dilemma: China's Effort to Protect its Energy Supply". *Strategic Analysis* 31(3): 467–489.

Jordan, Matt, Dawn Manley, Valerie Peters, and Ron Stoltz. 2012. *The Goals of Energy Policy: Professional Perspectives on Energy Security, Economics, and the Environment*. Albuquerque, NM: Sandia National Laboratories/U.S. Department of Energy.

Joshi, Ketan. 2014. "Why that Guy You Know Hates Renewable Energy". *Limited News*. From http://limitednews.com.au/2014/02/why-that-guy-you-know-hates-renewable-energy/, 13 February 2014.

Jowsey, Ernie. "Economic Aspects of Natural Resource Exploitation", *International Journal of Sustainable Development & World Ecology* 16(5): 303–307.

Kalicki, Jan H. and David L. Goldwyn. 2005. *Energy and Security: Toward a New Foreign Policy Strategy*. Baltimore: The Johns Hopkins University Press.

Kaplan, Robert D. 2005. "How We Would Fight China". *The Atlantic*, June.

Kashi, David. 2014. "NATO Countrie's 2013 Defense Spending Estimates are Likely To Upset US Officials". *International Business Times* Etienne Uzac, 25 February 2014.

Kazer, William. 2013. "China Forecasts 7.6% Economic Growth in 2013". *The Wall Street Journal*, 26 December.

Kendall, Timothy. 2005. *Ways of Seeing China: From the Yellow Peril to Shangrila*. Freemantle: Curtin University Books.

Kerr, David. 2012. "China and Inner Asia: New Frontiers and New Challenges". *The Forum: Discussing International Affairs and Economics* 2012(Summer): 21–28.

Kerr, David. 1999. "The Chinese and Russian Energy Sectors: Comparative Change and Potential Interaction". *Post-Communist Economies* 11(3): 337–372.

Khan, Joseph. 2005. "A Deft Balance in Orchestrating China's Oil Offer". *New York Times*, 7 July.

Khan, Joseph. 2005. "Behind China's Bid for Unocal: A Costly Quest for Energy Control". *New York Times*, 24 June.

Kil, Sang Hea. 2012. "Fearing Yellow, Imagining White: Media Analysis of the Chinese Exclusion Act of 1882". *Social Identities: Journal for the Study of Race, Nation and Culture* 18(6): 663–677.

Kilpatrick, Carolyn. 2005. "151 Cong. Rec. 11 (28 June 2005 to 13 July 2005)", *United States Government Printing Office*, Washington, 2005.

Koschorke, Klaus, Frieder Ludwig, Mariano Delgado, and Roland Spielsgart. 2007. *A History of Christianity in Asia, Africa, and Latin America, 1450–1990: A Documentary Sourcebook*. Grand Rapids, MI: Eerdmans.

Kristof, Nicholas D. 1989. "China Erupts … The Reasons Why". *New York Times*. 4 June.

Kristof, Nicholas D. 1993. "China Sees 'Market-Leninism' as Way to Future". *New York Times*, 6 September.

Kristof, Nicholas D. 1989. "Crackdown in Beijing; Troops Attack and Crush Beijing Protest; Thousands Fight Back, Scores are Killed". *New York Times*. 4 June.

Kruyt, Bert, D.P. van Vuuren, H.J.M. de Vries, and H. Groenenberg. 2009. "Indicators for Energy Security". *Energy Policy* 37: 2166–2181.

Lane, Charles. 2012. "Liberals' Green-Energy Contradictions". *Washington Post*, 15 October.

Langer, Gary. 2013. "Poll Finds Vast Gaps in Basic Views on Gender, Race, Religion and Politics". *ABC News*. From http://abcnews.go.com/blogs/politics/2013/10/polll-finds-vast-gaps-in-basic-views-on-gender-race-religion-and-politics/, 28 October 2013.

Lawson, George. 2012. "The Eternal Divide? History and International Relations". *European Journal of International Relations* 18: 203–226.

Lee, Don and Elizabeth Douglass. 2005. "Chinese Drop Takeover Bid for Unocal". *Los Angeles Times*, 3 August.

Legro, Jeffrey W. "What China Will Want: The Future Intentions of a Rising Power", *Perspectives on Politics* 5(3), September 2007: 515–534, 515.

Linklater, Andrew. 2005. "Political Community and Human Security". In *Critical Security Studies and World Politics*, edited by Ken Booth. Boulder: Colorado, Lynne Rienner Publishers, Inc., 113–133.

Litterick, David. 2005. "China Angered by US Oil Sale Intervention, Beijing Calls on Washington to Stop 'Interfering' in Free Enterprise". *Daily Telegraph*, 5 July.

Lloyd-Smith, Jake. 2005. "Anti-China Rhetoric as Unocal Ponders Offers". *Evening Standard*, 15 July.

Lohr, Steve. 2005. "The Big Tug of War Over Unocal". *New York Times*, 6 July.

Lohr, Steve. 2005. "Unocal Bid Denounced at Hearing". *New York Times*, 14 July.

Lohr, Steve. 2005. "Unocal Bid Opens Up New Issues of Security". *New York Times*, 13 July.

Lohr, Steve. 2005. "Who's Afraid of China Inc?" *New York Times*, 24 July.

Loschel, Andreas, Ulf Moslener, and Dirk T.G. Rubbelke. 2010. "Indicators of Energy Security in Industrialised Countries". *Energy Policy* 38: 1665–1671.

Luft, Gal. 2004. "Fueling the Dragon: China's Race Into the Oil Market". [Online] *Institute for the Analysis of Global Security*, Retrieved 10 February 2014, from www.iags.org/china.htm.

Macartney, George. 2000. In *Sinophiles and Sinophobes: Western Views of China*, edited by Colin Mackerras. Oxford: Oxford University Press.

Mackerras, Colin. 2000. *Sinophiles and Sinophobes: Western Views of China*. Oxford: Oxford University Press.

Marchick, David, Mark Plotkin, and David Fagan. 2005. "National Security Regulation of Foreign Investments and Acquisitions in the United States". *China Law & Practice*, June.

Martin, W.F. and E.M. Harrje, "The International Energy Agency", in Energy and Security: Toward a New Foreign Policy Strategy, edited by J.H. Kalicki., D.L. Goldwyn. Woodrow Wilson Press, Washington.

McGivering, Jill. 2006. "Three Gorges Dam's Social Impact". [Online] *BBC News*. Retrieved 18 March 2014, from http://news.bbc.co.uk/1/hi/world/asia-pacific/5000198.stm.

McGregor, James. 2005. "Advantage, China". *Washington Post*. 31 July.

McKenna, Bernard. 2004. "Critical Discourse Studies: Where To From Here?" *Critical Discourse Studies* 1(1): 9–39.

McMillan, Andrew Frew. 2011. "China's Role as 'World's Factory' Coming to an End". *Invest China – Special Report*. From www.cnbc.com/id/41035650, 6 February 2011.

Mearsheimer, J. John. 2010. "The Gathering Storm: China's Challenge to US Power in Asia". *The Chinese Journal of International Politics* 3: 381–396.

Mearsheimer, John. 2005. "The Rise of China Will Not Be Peaceful at All". *The Australian*. 18 November.

Mendoza, Juan Gonzales de. 1853. *The History of the Great and Mighty Kingdom of China and the Situation Thereof*. Translated by Sir George T. Staunton. London: The Hakluyt Society.

Mingyuan, Wang. 2005. "Government Incentives to Promote Renewable Energy in the United States". *Temple Journal of Science, Technology & Environmental Law* 24(1): 355–366.

Mirsky, Jonathan. 1972. "China after Nixon". *Annals of the American Academy of Political and Social Science* 402(China in the World Today [Jul., 1972]): 83–96.

Mohitpour, M. 2008. *Energy Supply and Pipeline Transportation: Challenges & Opportunities*. New York, ASME.

Moms, Gijs. 2004. *Electric Vehicle: Technology and Expectations in the Automobile Age*. Baltimore: Johns Hopkins University Press.

Moore, John Frederick. 1998. "BP to Acquire Amoco". *CNN Money*, from http://money.cnn.com/1998/08/11/deals/bp/, 11 August 1998.

Morath, Eric. 2013. "Government's $421 Billion Bailout Turns Profitable". *The Wall Street Journal*, 21 November.

Morrison, Wayne M. 2005. "China–U.S. Trade Issues". *Congressional Research Service*. Washington, DC, 19 July.

Mortished, Carl. 2005. "China Eyes Europe After Unocal Rebuff". *The Times*, 3 August.

Mouawad, Jad. 2005. "Congress Calls for a Review of the Chinese Bid for Unocal". *New York Times*, 27 July.

Mouawad, Jad. 2005. "Foiled Bid Stirs Worry for U.S. Oil". *New York Times*, 11 August.

Mufson, Steven. 2013. "Does OPEC Still Have the U.S. Over a Barrel?" *Washington Post*, 11 October.

Mufson, Steven. 2008. "Oil Closes Over $100 for 1st Time". *Washington Post*. 20 February.

Mufson, Steven. 1999. "Zigzagging over China". *World Policy Journal*, 1999/2000: 97–103.

Najam, Adil, David Runnalls, and Mark Halle. 2007. *Environment and Globalization: Five Propositions*. Winnipeg, Manitoba: International Institute for Sustainable Development.

Nakashima, Ellen. 2011. "In a World of Cybertheft, U.S. Names China, Russia as Main Culprits". *Washington Post*, 3 November.

Nakashima, Ellen and William Wan. 2014. "U.S. Announces First Charges Against Foreign Country in Connection With Cyberspying". *Washington Post*. 19 May.

Nathan, Andrew J. and Andrew Scobell. 2012. "How China Sees America". *China–US Focus*. Retrieved 17 October 2013, from www.chinausfocus.com/foreign-policy/how-china-sees-america/.

Nanto, Dick K., James K. Jackson, Wayne M. Morrison, and Lawrence Kumins. 2005. "China and the CNOOC bid for Unocal: Issues for Congress". *Congressional Research Service*, Washington, DC: The Library of Congress.

Narodoslawsky, Michael, Anneliese Niederl-Schmidinger, and Laszlo Halasz. 2008. "Utilizing Renewable Resources Economically: New Challenges and Chances for Process Development". *Journal of Cleaner Production* 16: 164–170.

National Research Council. 2010. *Electricity from Renewable Resources: Status, Prospects, and Impediments*. Washington, DC: The National Academies Press.

National Security Strategy of the United States of America (NSS). 2002. *National Security Strategy of the United States of America*. The White House. Washington, DC.

Newport, Frank. 2009. "Women More Likely to Be Democrats, Regardless of Age". *Gallup*, from www.gallup.com/poll/120839/women-likely-democrats-regardless-age. aspx, 12 June 2009.

Ney, Bob. 2005. "151 Cong. Rec. 90, (daily ed. 30 June 2005)", *GPO.gov*, from www. gpo.gov/fdsys/pkg/CREC-2005-06-30/html/CREC-2005-06-30-pt1-PgH5570-2.htm.

Nixon, Richard. 1973. "Address to the Nation About National Energy Policy". In *The American Presidency Project*, edited by Gerhard Peters and John T. Woolley. From www.presidency.ucsb.edu/ws/?pid=4051, 25 November 1973.

Nixon, Richard. 1973. "Address to the Nation About Policies to Deal With the Energy Shortages". [Online] In *The American Presidency Project*, edited by Gerhard Peters and John T. Woolley. From www.presidency.ucsb.edu/ws/index.php?pid=4034&st=pr oject+independence&st1=, 7 November 1973.

Nixon, Richard. 1973. "Remarks About the Nation's Energy Policy". [Online] In *The American Presidency Project*, edited by Gerhard Peters and John T. Woolley. From www. presidency.ucsb.edu/ws/index.php?pid=3953&st=energy&st1=oil, 8 September 1973.

Nixon, Richard. 1973. "Remarks on Transmitting a Special Message to the Congress on Energy Policy". [Online] In *The American Presidency Project*, edited by Gerhard Peters and John T. Woolley. From www.presidency.ucsb.edu/ws/index.php?pid=3816 &st=outer+continental+shelf&st1=, 18 April 1973.

Nye, Joseph. 2005. "The Rise of China's Soft Power". *Wall Street Journal Asia*, 29 December.

Obama, Barak. 2011. "Remarks at Georgetown University". In *The American Presidency Project*, edited by Gerhard Peters and John T. Woolley. From www.presidency.ucsb. edu/ws/index.php?pid=90196&st=energy+security&st1=china, 30 March 2011.

Obama, Barak. 2009. "Remarks by the President on Jobs, Energy Independence, and Climate Change". In *The American Presidency Project*, edited by Gerhard Peters and John T. Woolley. Retrieved 22 September 2014, from www.presidency.ucsb.edu/ws/ index.php?pid=85689&st=&st1=, 04 October 2009.

OCHA. 2010. *Energy Security and Humanitarian Action: Key Emerging Trends and Challenges*. OCHA: Policy Development and Studies Branch, UN Office for the Coordination of Humanitarian Affairs (OCHA): 1–14.

O'Rourke, Ronald. 2013. "China Naval Modernization: Implications for U.S. Navy Capabilities – Background and Issues for Congress". *Congressional Research Service*. Washington, DC: 1–115.

O'Rourke, Ronald. 2014. "China Naval Modernization: Implications for U.S. Navy Capabilities – Background and Issues for Congress". *Congressional Research Service*. Washington, DC.

Ou, Hsin-yun. 2010. "Mark Twain's Racial Ideologies and His Portrayal of the Chinese". *Concentric: Literary and Cultural Studies* 36(2): 33–59.

Paish, Oliver. 2002. "Small Hydro Power: Technology and Current Status". *Renewable and Sustainable Energy Reviews* 6: 537–556.

Pan, Chengxin. 2012. *Knowledge, Desire and Power in Global Politics: Western Representations of China's Rise*. Cheltenham, UK: Edward Elgar Publishing Limited.

Peerenboom, Randall. 2008. *China Modernizes: Threat to the West or Model for the Rest?* Oxford: Oxford University Press.

Pelosi, Nancy. 2005. "Pelosi Statement on Amendment to Block Chinese Bid to Acquire Unocal". *Pelosi.house.gov*. From http://pelosi.house.gov/news/press-releases/pelosi-statement-on-amendment-to-block-chinese-bid-to-acquire-unocal, 30 June 2005.

Pendergrast, Mark. 1999. *Uncommon Grounds: The History of Coffee and How it Transformed Our World*. New York: Basic Books.

Perlez, Jane. 2012. "Continuing Buildup, China Boosts Military Spending More Than 11 Percent". *International New York Times*, 4 March.

Pew Research Center. 2015. *Global Indicators Database*. Retrieved 26 June 2015 from www.pewglobal.org/database/indicator/33/survey/17/.

Polo, Marco. 1997. *The Travels of Marco Polo*. Ware, Hertfordshire: Wordsworth Editions Limited.

Pombo, Richard. 2005. 151 Cong. Rec. 11 (28 June 2005 to 13 July 2005), Available from *United States Government Printing Office*, Washington, 2005.

Prozorov, Sergei. 2011. "The Other as Past and Present: Beyond the Logic of 'Temporal Othering' in IR Theory". *Review of International Studies* 37(3): 1273–1293.

Reagan, Ronald. 1982. "Remarks at the Opening Ceremonies for the Knoxville International Energy Exposition (World's Fair) in Tennessee". [Online] In *The American Presidency Project*, edited by Gerhard Peters and John T. Woolley. From www.presidency.ucsb.edu/ws/index.php?pid=42470&st=Remarks+at+the+Opening+Ceremonies+for+the+Knoxville+International+Energy+Exposition+%28World%5C%27s+Fair%29+in+Tennessee&st1=.

Red State. 2008. "Democrats and Sissies". *Red State*. Retrieved 05 October 2013, from http://archive.redstate.com/stories/the_parties/democrats/democrats_and_sissies.

Regnier, Eva. 2007. "Oil and Energy Price Volatility". *Energy Economics* 29: 405–427.

Resurrección, Bernadette P. 2013. "Persistent Women and Environmental Linkages in Climate Change and Sustainable Development Agendas". *Women's Studies International Forum* 40: 33–43.

Reus-Smit, Chrisian. 2012. "International Relations, Irrelevant? Don't Blame Theory". *Millennium – Journal of International Studies* 40: 525–540.

Reynolds, Thomas. 2005. "Reynolds Challenges China's Bid for Unocal". *Project Vote Smart*. Retrieved 12 October 2013, from http://votesmart.org/public-statement/111630/#.UpDqMsQ73o8.

Rivera Brooks, Nancy. 1999. "BP Amoco Will Acquire Arco for $27 Billion". *Los Angeles Times*, 1 April.

Roberts, Paul. 2004. *The End of Oil*. London: Bloomsbury.

Roby, Martha. 2005. "159 Cong. Rec. 92, (daily ed. 25 June 2013)", *GPO.gov*, from www.gpo.gov/fdsys/granule/CREC-2013-06-25/CREC-2013-06-25-pt1-PgH4015/content-detail.html.

Rocha, Euan. 2013. "CNOOC Closes $15.1 Billion Acquisition of Canada's Nexen". *Reuters*, 25 February.

Roff, Peter. 2013. "The Utterly Pointless War on Coal". [Online] *U.S. News*, Retrieved 2 February 2014, from www.usnews.com/opinion/blogs/peter-roff/2013/10/29/obamas-war-on-coal-runs-counter-to-global-energy-trends, 29 October 2013.

Rowe, William T. 2009. *China's Last Empire: The Great Qing*. Cambridge, MA: The Belknap Press of Harvard University Press.

Roy, Denny. 1996. "The "China Threat" Issue: Major Arguments". *Asian Survey* 36(8): 13.

Rumelili, Bahar. 2004. "Constructing Identity and Relating to Difference: Understanding the EU's Mode of Differentiation". *Review of International Studies* 30(1): 27–47.

Salpukas, Agis. 1999. "It's Official: BP Is Planning To Buy ARCO". *New York Times*, 2 April.

Samaras, Constantine and Henry H. Willis. 2013. *Capabilities-Based Planning for Security at Department of Defense Installations*. Santa Monica, CA: Homeland Security and Defense Center (RAND).

Samuelson, Robert J. 2005. "China's Oil Bid: A Battle to Avoid..". *Washington Post*, 6 June.

Samuelson, Robert J. 2008. "The Real China Threat". *Washington Post*, 20 August.

Scharff, Virginia. 1991. *Taking the Wheel: Women and the Coming of the Motor Age.* New York: Free Press.

Shlapak, David A., David T. Orletsky, Toy I. Reid, Murray Scott Tanner, and Barry Wilson. 2009. *A Question of Balance: Political Context and Military Aspects of the China-Taiwan Dispute.* Santa Monica, CA: RAND Corporation.

Schortgen, Francis. 2006. "Protectionist Capitalists vs. Capitalist Communists: CNOOC's Failed Unocal Bid In Perspective". *Asia Pacific: Perspectives* VI(2): 2–11.

Schrecker, Ellen W. 1988. "Archival Sources for the Study of McCarthyism". *The Journal of American History* 75(1): 197–208.

Schuman, Michael. 2013. "The Chinese Communist Party's Biggest Obstacle Is the Chinese Communist Party". *Time*, 25 November.

Schwartz, Nelson D. 2004. "Inside the Head of BP" *Fortune Magazine*, 26 July.

Securing America's Future Energy (SAFE). 2012. *The New American Oil Boom: Implications for Energy Security*, Washington, DC: Securing America's Future Energy.

Sehgal, Amrish. 2003. "China and the Doctrine of Asymmetrical Warfare". *Bharat Rakshak Monitor* 6(1).

Seshagiri, Urmila. 2006. "Modernity's (Yellow) Perils: Dr. Fu-Manchu and English Race Paranoia". *Cultural Critique* 62(Winter): 162–194.

Shambaugh, David. 2013. *China Goes Global: The Partial Power.* Oxford: Oxford University Press.

Shanker, Thom. 2012. "Pentagon Tries to Counter Cheap, Potent Weapons". *New York Times*, 9 January.

Sharan, Sunil. 2011. "America is Losing the Green Energy Race". *Washington Post*, 7 December.

Shaw, Han-Yu. 2012. "The Inconvenient Truth Behind the Diaoyu/Senkaku Islands". *International New York Times*, 19 September.

"Shell and Texaco 'Merger Talks'". 1998. *BBC Online Network*. From http://news.bbc.co.uk/1/hi/business/127244.stm, 6 July 1998.

Shih, David. 2009. "The Color of Fu-Manchu: Orientalist Method in the Novels of Sax Rohmer". *The Journal of Popular Culture* 42(2): 304–317.

Shirk, Susan L. 2007. *China: Fragile Superpower.* Oxford: Oxford University Press.

Sieminski, Adam E. 2005. "World Energy Futures". In *Energy and Security: Toward a New Foreign Policy Strategy*, edited by Jan H. Kalicki and David L. Goldwyn. Baltimore: The Johns Hopkins University Press.

Siklos, Richard. 2005. "Sell to the Chinese if Their Money is Green". *Sunday Telegraph*, 10 July.

Singer, Clifford. 2008. "Oil and Security". *Policy Analysis Brief.* Muscatine, IA: The Stanley Foundation.

Sklar, Scott. 2012. "American Exceptionalism and Renewable Energy: What the Tea Party Missed in 2011". *RenewableEnergyWorld.com*, from www.renewableenergy-world.com/rea/news/article/2012/01/American-exceptionalism-and-renewable-energy-what-the-tea-party-missed-in-2011, 4 January 2012.

Skonieczny, Amy. 2001. "Constructing NAFTA: Myth, Representation, and the Discursive Construction of U.S. Foreign Policy". *International Studies Quarterly* 45: 433–454.

Slivinski, Stephen. 2007. "The Corporate Welfare State: How the Federal Government Subsidizes U.S. Businesses". *CATO Institute*, Washington, DC.

Smith, Adam. 1776. *The Wealth of Nations*. Petersfield: Harriman House Ltd.

Smith, James L. 2012. "On the Portents of Peak Oil (And Other Indicators of Resource Scarcity)". *Energy Policy* 44: 68–78.

Smith, Karl. 2013. "Will Natural Gas Stay Cheap Enough to Replace Coal and Lower US Carbon Emissions". *Forbes*, 22 March.

Sovacool, Benjamin K. 2009. "The Importance of Comprehensiveness in Renewable Electricity and Energy Efficiency Policy". *Energy Policy* 37: 1529–1541.

Spegele, Roger D. 1992. "Richard Ashley's Discourse for International Relations". *Millennium – Journal of International Studies* 21: 147–182.

Spence, Jonathan D. 1998. *The Chan's Great Continent: China in Western Minds*. New York: W. W. Norton & Company.

Spencer, Richard. 2005. "China's US Ambitions Thwarted as Two Major Bids Scuppered". *The Daily Telegraph*, 21 July.

Steele, A. T. 1966. *The American People and China*. New York: The McGraw-Hill Book Company.

Stevens, Paul. 2008. *The Coming Oil Supply Crunch*. London: Royal Institute for International Affairs.

Stevens, Paul. 2005. "Oil Markets". *Oxford Review of Economic Policy* 21(1): 19–42.

Suro, Roberto. 1999. "Reno's Upset With Belated Video Disclosure". *Washington Post*. 3 September.

Tarling, Nicholas. 1967. *China and Its Place in the World*. Auckland: Blackwood & Janet Paul Ltd.

Taylor, Jerry. 2005. "CNOOC Bid for Unocal No Threat to Energy Security". [Online] *CATO Institute*, Retrieved 30 March 2014, from www.cato.org/publications/free-trade-bulletin/cnooc-bid-unocal-no-threat-energy-security, 19 July 2005.

Tekin, Beyza Ç. 2010. *Representations and Othering in Discourse: The Construction of Turkey in the EU Context*. Amsterdam: John Benjamins Publishing Company.

Thompson, Loren. 2014. "Five Reasons China Won't Be A Big Threat To America's Global Power", [Online], *Forbes*, Retrieved 27 March 2015, from www.forbes.com/sites/lorenthompson/2014/06/06/five-reasons-china-wont-be-a-big-threat-to-americas-global-power/, 6 June 2014.

Tie, Li. 2012. "Whitewashing the Three Gorges Dam". [Online] *PROBE International*, Retrieved 18 March 2014, from http://journal.probeinternational.org/2012/10/15/whitewashing-the-three-gorges-dam/.

Tippee, Bob. 2012 "Defining Energy Security". *Oil & Gas Journal*. 23 January.

Tobak, Steve. 2013. "Don't Buy an Electric Car". *FOX Business*. Retrieved 5 November 2013, from www.foxbusiness.com/business-leaders/2013/04/05/dont-buy-electric-car/?intcmp=sem_outloud.

Trigault, Nicolas. 1942. *The China that Was: China as Discovered by the Jesuits at the Close of the Sixteenth Century*. Milwaukee: Bruce Publishing Co.

Tschang, Chi-Chu. 2008. "What Bud's Takeover Means for China". *BloombergBusinessweek*, 14 July.

Turner, Louis. 1983. "OPEC". In *The Third Oil Shock: The Effects of Lower Oil Prices*, edited by Joan Pearce. London, Routledge & Kegan Paul Ltd.: 82–90.

Turner, Oliver. (2011). "Sino-US Relations Then and Now: Discourse, Images, Policy". *Political Perspectives* 5(3): 27–45.

Udall, Tom. 2005. *Understanding the Peak Oil Theory: Hearing Before the Subcommittee on Energy and Air Quality*. Committee of Energy and Commerce. Washington, DC, U.S. Government Printing Office.

US Congress, U.S.–China Economic and Security Review Commission. 2005 *2005 Report to Congress of the U.S.–China Economic and Security Review Commission.* (109th Congress, 1st Session). Washington, U.S. Government Printing Office, 2005.

US Cong. Senate. 2005. *To Prohibit the Merger, Acquisition, or Takeover of Unocal by CNOOC Ltd. of China.* 15 July 2005. 109th Cong., 1st sess. S.1412. [Online] Washington, Government Printing Office, Available from GPO.gov, www.gpo.gov/fdsys/pkg/BILLS-109s1412is/pdf/BILLS-109s1412is.pdf.

US Department of Energy. (DOE). 2013. "Renewable Electricity Generation". [Online] *Energy.gov: Office of Energy Efficiency & Renewable Energy.* Retrieved 18 July 2011 from http://energy.gov/eere/renewables.

US House, Committee on Energy and Commerce. 2001. *National Energy Policy: Crude Oil and Refined Petroleum Products.* Hearing, 30 March 2001, (Serial No. 107–12). Washington, Government Printing Office, 2001.

US House, Committee on Energy and Commerce. 2005. *Understanding the Peak Oil Theory.* Hearing, 7 December 2005 (Serial No. 109–41). Washington, Government Printing Office, 2005.

US House, Committee on Finance. 1996. *China Most-Favored-Nation (MFN) Status.* Hearing, 6 June 1996 (S. Hrg. 104–871). Washington: Government Printing Office, 1997.

Voltaire. 1766. *The Philosophy of History.* New York: The Citadel Press.

Voltaire. 2000. "Splendid Secular Governance". In *Sinophiles and Sinophobes: Western Views of China,* edited by Colin Mackerras. Oxford: Oxford University Press, 35–40.

Vukovich, Daniel F. 2012. *China and Orientalism: Western Knowledge Production and the P.R.C.* New York: Routledge.

Vuori, Juha. A. 2008. "Illocutionary Logic and Strands of Securitization: Applying the Theory of Securitization to the Study of Non-Democratic Political Orders". *European Journal of International Relations* 14: 65–99.

Waever, Ole. 1998. "Insecurity, Security, and Asecurity in the West European Non-War Community". In *Security Communities,* edited by Emanuel Adler and Michael Barnett. Cambridge: Cambridge University Press, 69–119.

Waever, Ole. 2011. "Politics, Security, Theory". *Security Dialogue* 42(4–5): 465–480.

Walker, R.B.J. 1993. *Inside/Outside: International Relations as Political Theory.* Cambridge: Cambridge University Press.

Walter, Richard and Benjamin Robins. 1974. *A Voyage Round the World in the Years MDCCXL, I, II, II, IV by George Anson.* London: Oxford University Press.

Walter, Richard and Benjamin Robins. 2000. "Jesuitical Fictions". In *Sinophiles and Sinophobes: Western Views of China,* edited by Colin Mackerras. Oxford: Oxford University Press, 47–49.

Wan, William. 2014. "As Budgets Soar, China Still Fears Its Military Isn't Growing Fast Enough". *Washington Post,* 7 March.

Wan, William. 2013. "Witnesses to Tiananmen Square Struggle With What to Tell Their Children". *Washington Post,* 2 June.

Wang, Yuan-Kang. 2004. "Offensive Realism and the Rise of China". *Issues & Studies* 40(1): 173–201.

Warner, Geoffrey. 2007. "Nixon, Kissinger and the rapprochement with China, 1969–1972". *International Affairs* 83(4): 763–781.

Washington Institute. 1991. "Policy Analysis: A Post-Gulf War Assessment". [Online] *The Washington Institute for Near East Policy,* Retrieved 30 March 2014, from www.washingtoninstitute.org/policy-analysis/view/a-postgulf-war-assessment.

Wayne, Leslie and David Barboza. 2005. "Unocal Deal: A Lot More Than Money is At Issue". *New York Times*, 24 June.

Weisman, Jonathan. 2005. "In Washington, Chevron Works to Scuttle Chinese Bid". *Washington Post*, 16 July.

Wendt, Alexander. 1987. "The Agent-Structure Problem in International Relations Theory". *International Organization* 41(3): 335–370.

Wendt, Alexander. 1999. *Social Theory of International Politics*. Cambridge: Cambridge University Press.

Whoriskey, Peter. 2011. "Conventional Gas-Powered Cars Starting to Match Hybrids in Fuel Efficiency". *Washington Post*, 9 March.

"Why China's Unocal Bid Ran Out of Gas". 2005. *BloombergBusinessweek*. Retrieved 8 September 2013, from www.businessweek.com/stories/2005-08-03/why-chinas-unocal-bid-ran-out-of-gas.

"Why is Renewable Energy So Expensive?" 2014. *The Economist*, 5 January.

Wines, Michael. 2011. "China Admits Problems With Three Gorges Dam". *New York Times*, 19 May.

Wines, Michael and Edward Wong. 2011. "China's Push to Modernize Military Is Bearing Fruit". *New York Times*, 5 January.

Winter, Nicholas J. G. 2010. "Masculine Republicans and Feminine Democrats: Gender and Americans' Explicit and Implicit Images of the Political Parties". *Political Behavior* 34(2): 587–618.

Winzer, Christian. 2012. "Conceptualizing Energy Security". *Energy Policy* 46: 36–48.

Wolfers, Arnold. 1962. *Discord and Collaboration; Essays on International Politics*. Baltimore: Johns Hopkins Press.

Wolfers, Arnold. 1952. " 'National Security' as an Ambiguous Symbol". *Political Science Quarterly* 67(4): 481–502.

woodyi...@my-deja.com. 1999. "Wen Ho Lee should be SHOT!!". [Online] *Google Groups*. Retrieved 12 November 2013, from https://groups.google.com/forum/#!topic/alt.politics/tl6KRa8VK5o, 31 December 1999.

World Bank. 2014. "China Overview". *The World Bank*. Retrieved 28 May 2014, from, www.worldbank.org/en/country/china/overview.

World Nuclear Association. 2015. "Renewable Energy and Electricity". World Nuclear Association, Retrieved 18 March 2015, from www.world-nuclear.org/info/Energy-and-Environment/Renewable-Energy-and-Electricity/.

Worstall, Tim. 2011. "Chuck Schumer and the China Currency Bill". *Forbes*, 16 October.

Wu, William F. 1982. *The Yellow Peril: Chinese Americans in American Fiction 1850–1940*. Hamden, CT: Archon Books.

Wustenhagen, Rolf, Maarten Wolsink, and Mary Jean Burer. 2007. "Social Acceptance of Renewable Energy Innovation: An Introduction to the Concept". *Energy Policy* 35: 2683–2691.

Xia, Ming. 2006. " 'China Threat' or a 'Peaceful Rise of China'?", [Online], *New York Times*, Retrieved 30 March 2015, from www.nytimes.com/ref/college/coll-china-politics-007.html.

Yergin, Daniel. 1993. *The Prize: The Epic Quest for Oil, Money and Power*. London: Pocket Books.

Yergin, Daniel. 2005. "Energy Security and Markets". In *Energy and Security: Toward a New Foreign Policy Strategy*, edited by Jan H. Kalicki and David L. Goldwyn. Baltimore: The Johns Hopkins University Press.

Yergin, Daniel. 2006. "Ensuring Energy Security". *Foreign Affairs* 85(2): 69–82.

Yongping, Feng. 2006. "The Peaceful Transition of Power from the UK to the US". *Chinese Journal of International Politics* 1: 83–108.

Young, Alf. 2005. "US Would be Wise to Oil Wheels of China's Progress". *The Herald.* 28 June.

Zichal, Heather. (2012). "Increasing Energy Security". [Online] *U.S. Department of Energy*, Retrieved 17 October 2013, from http://energy.gov/articles/increasing-energy-security.

Index

Page numbers in *italics* denote tables, those in **bold** denote figures.